Turbulence and Flow–Sediment Interactions in Open-Channel Flows

Turbulence and Flow–Sediment Interactions in Open-Channel Flows

Editor

Roberto Gaudio

MDPI • Basel • Beijing • Wuhan • Barcelona • Belgrade • Manchester • Tokyo • Cluj • Tianjin

Editor
Roberto Gaudio
Università della Calabria
Italy

Editorial Office
MDPI
St. Alban-Anlage 66
4052 Basel, Switzerland

This is a reprint of articles from the Special Issue published online in the open access journal *Water* (ISSN 2073-4441) (available at: https://www.mdpi.com/journal/water/special_issues/Turbulence_Flow_Sediment).

For citation purposes, cite each article independently as indicated on the article page online and as indicated below:

LastName, A.A.; LastName, B.B.; LastName, C.C. Article Title. *Journal Name* **Year**, *Volume Number*, Page Range.

ISBN 978-3-03943-899-0 (Hbk)
ISBN 978-3-03943-900-3 (PDF)

Contents

About the Editor

Roberto Gaudio

Education

Ph.D. in Hydraulic Engineering for Environment and Territory at the University of Calabria.
Master degree cum laude in Civil Engineering (Hydraulics), at the University of Calabria.

Actual Position

Full Professor (Hydraulics) at the Dept. of Civil Engineering, University of Calabria, Italy.
Head of the Department of Civil Engineering, University of Calabria, Italy.
Prof. of Hydraulics and Fluvial Hydraulics (Bachelor and Master in Civil Engineering).

Research activity

He has co-authored about 150 publications on fluvial hydraulics and urban drainage.

Research interests:

– Floods.
– Sediment transport, erosion, deposition.
– Local scouring at structures (bed sills, bridge piers and abutments).
– Turbulence and water–sediment interactions.
– Small-scale physical hydraulic models.

Research projects

Coordinator or partner in many national and international research projects (funded by the Italian Ministry or by the Commission of the European Communities, Directorate General for Science, Research and Development), carried out in Italy or at the Hydraulics Laboratory of HR Wallingford Ltd., Oxfordshire, UK.

Editorial

Turbulence and Flow–Sediment Interactions in Open-Channel Flows

Roberto Gaudio

Dipartimento di Ingegneria Civile, Università della Calabria, 87036 Rende, Italy; gaudio@unical.it

Received: 26 October 2020; Accepted: 12 November 2020; Published: 13 November 2020

Abstract: The main focus of this Special Issue of *Water* is the state-of-the-art and recent research on turbulence and flow–sediment interactions in open-channel flows. Our knowledge of river hydraulics is becoming deeper and deeper, thanks to both laboratory/field experiments related to the characteristics of turbulence and their link to the erosion, transport, deposition, and local scouring phenomena. Collaboration among engineers, physicists, and other experts is increasing and furnishing new inter/multidisciplinary perspectives to the research in river hydraulics and fluid mechanics. At the same time, the development of both sophisticated laboratory instrumentation and computing skills is giving rise to excellent experimental–numerical comparative studies. Thus, this Special Issue, with ten papers by researchers from many institutions around the world, aims at offering a modern panoramic view on all the above aspects to the vast audience of river researchers.

Keywords: turbulence; local scouring; erosion; transport; deposition; open-channel flows; laboratory experiments

---

1. Introduction to the Special Issue

The investigation of the interactions between flow and sediments in natural bed streams is one of the most fascinating research topic in the field of fluvial hydraulics, especially if the issues to be addressed are focused on the role of turbulence and its main characteristics in the sediment erosion, deposition and transport processes [1–5], as well as in the presence of vegetation [6]. Similarly, a comprehensive understanding of the flow dynamics in the near-bed flow zone is crucial for many different practical applications, such as bridge pier scour, pipeline scour, scour at bed sills, and other situations in which anthropization affects the behavior of water flow, turbulence, and sediment transport [7–10].

The Special Issue comprises ten original papers on turbulence and flow–sediment interactions in open-channel flows, which can be divided into two main categories: experimental [1–4,6–8,10] and field [5,9] studies. At the same time, among the experimental studies, it is possible to make another distinction on the basis of the instrumentation adopted. Indeed, two different sophisticated technologies were used to capture the instantaneous velocity field: the first one is based upon the Doppler shift effect (Acoustic Doppler Velocimeter, ADV) [2,6–8] and the second is an optical method of flow visualization (Particle Image Velocimetry, PIV) [1,3,4,10]. As regards the field studies [5,9], the Acoustic Doppler Current Profiler (ADCP) was employed, which is based on the Doppler effect as well as the ADV. From the viewpoint of turbulence analyses, all the contributions present in-depth and detailed statistical investigations. In fact, starting from the velocity fields, the researches were extended to the examination of vorticity, viscous and Reynolds stresses, turbulence indicators, turbulence intensity, turbulent length scales, turbulent kinetic energy and turbulent kinetic energy budget, and even to the study of wavelet spectrum, quadrant analysis and Reynolds stress anisotropy.

Thus, in order to provide a view of the key points of the contributions to this Special Issue, the papers were considered together and summarized in the following Section, according to the alphabetical order of the lead author.

2. Overview of the Contributions of the Special Issue

2.1. Kinematics of Particles at Entrainment and Disentrainment (Aleixo et al., 2020)

The objective of this study was the characterization of entrainment and disentrainment of sediment particles of uniform granular beds in turbulent open-channel flows, by performing laboratory experiments conducted by using a PIV system. The Authors demonstrated that sediment entrainment occurred at a wide range of turbulent flow velocities, with a prevalence of sweep and outward interactions. Specifically, four types of particle entrainment were identified: (i) that caused by hydrodynamic forces (Type A); (ii) that promoted by other sediments rolling or saltating near the particle at rest (Type B); (iii) that due to direct collisions between particles (Type C); and (iv) that due to simultaneous pickup of several sediments (Type D). As regards disentrainment events, it was found that negative values of instantaneous velocity fluctuations in streamwise direction were observed for the whole sample of particles. It was also revealed that the bed topography influenced the disentrainment events: sediments in motion tend to be trapped within pockets if bed depressions are found.

2.2. Hydrodynamic Structure with Scour Hole Downstream of Bed Sills (Ben Meftah et al., 2020)

In this study, experimental measurements of the scour hole downstream of bed sills with non-cohesive sediments was investigated. The flow field was measured within the equilibrium scour hole using an ADV. It was found that the flow in the scour hole can be characterized by three distinct regions: (i) a free entering jet flow; (ii) a second region located near the scour bottom, extending upstream owing to eddies generated by the jet diffusion; and (iii) a third less-turbulent region, localized downstream and characterized by an almost unidirectional flow in the streamwise direction. Furthermore, the phenomenological theory of turbulence was applied to predict the maximum equilibrium scour depth. With this approach, a new scaling of the maximum scour depth at equilibrium was obtained, which was validated using the experimental data.

2.3. Flow–Sediment Turbulent Ejections: Interaction between Surface and Subsurface Flow in Gravel-Bed Contaminated by Fine Sediment (Bustamante-Penagos and Niño, 2020)

This study is related to experiments on a surface alluvial stream polluted with fine sediment percolated into the bed. PIV measurements were performed and velocity data were analyzed by scatter plots, power spectra and wavelet analysis of turbulent fluctuations, finding changes with and without the presence of these fine deposits. Specifically, the results revealed that the sediment ejections change the patterns of turbulent structures and the distribution of the turbulence interactions, implying that the flow does not have a typical rough-wall open-channel flow turbulence. Additionally, the sediment ejections increase the energy both in the production zone and inertial subrange.

2.4. Wavelet Coherency Structure in Open Channel Flow (Chen et al., 2019)

In this study, based on PIV data, the wavelet coherency analysis was applied to catch the coherent structures in a steady open-channel flow. As a result, it was demonstrated that the high value peaks in the pre-multiplied wavelet power spectrum curves stand for the energetic scales in the signal, and the high value areas in the local wavelet spectrum give both the scales and the time instants of energetic motions. The methodology can also detect the scale and the occurrence time instants of energetic motions and the inner structure of them. Furthermore, it was found that the wavelet coherency analysis supports the hairpin packets model in open-channel flows.

2.5. Bedform Morphology in the Area of the Confluence of the Negro and Solimões-Amazon Rivers, Brazil (Gualtieri et al., 2020)

The aim of this study was the investigation of the bedforms observed in the area of the confluence of the Negro and Solimões/Amazon Rivers (Brazil), whose morphology was acquired with an Acoustic Doppler Current Profiler (ADCP). The results showed that wavelength and wave height of the bedforms increased as the river discharge increased. Furthermore, the dunes were characterized by low-angles and, while several dunes were in equilibrium with the flow, several largest bedforms were found to be probably adapting to discharge changes in the river.

2.6. Man-Induced Discrete Freshwater Discharge and Changes in Flow Structure and Bottom Turbulence in Altered Yeongsan Estuary, Korea (Kang and Lee, 2020)

Using ADCP measurements in the Yeongsan estuary (Korea), this study aimed at examining the flow field resulting from the opening of the dam gate and the release of water with different properties, such as salinity, temperature and flow rate. Specifically, comparisons between the bottom turbulent kinetic energy (TKE) and the suspended sediment concentration (SSC) were performed. As a result, it was demonstrated that the surface freshwater discharge from the dam gate affects the behavior of water flow, bottom turbulence and sediment transport in the study area.

2.7. Turbulence Characteristics before and after Scour Upstream of a Scaled-Down Bridge Pier Model (Lee and Hong, 2019)

This study aims at understanding the near-bed turbulence characteristics and the resulting sediment transport around a pier, by using an ADV in laboratory experiments on scaled-down bridge pier models. Velocities and turbulence intensities as well as bed elevations before and after the scour were measured with an ADV. The results show that the mean flow variables are not sufficient to characterize the complex turbulent flow field around the pier leading to the maximum scour, because of unsteady flows. Instead, the quadrant analysis of velocity fluctuations revealed that bursts and sweeps are the primary forcing function for creating the scour hole at initial stage.

2.8. Turbulent Flow Field around Horizontal Cylinders with Scour Hole (Penna et al., 2020a)

This study presents the results of an experimental investigation on scoured horizontal cylinders, varying the gap between the cylinder and the bed surface. A PIV system was used to measure the flow field in a vertical plane at the end of the scouring process. The results revealed that suspended and laid on cylinders behave differently from half-buried cylinders if subjected to the same hydraulic conditions. In the latter case, vortex shedding downstream of the cylinder is suppressed by the presence of the bed surface that causes an asymmetry in the development of the vortices. This implies that strong turbulent mixing processes occur downstream of the uncovered cylinders, whereas in the case of half-buried cylinders they are confined within the scour hole.

2.9. Anisotropy in the Free Stream Region of Turbulent Flows through Emergent Rigid Vegetation on Rough Beds (Penna et al., 2020b)

In this study, an experimental investigation, based on ADV measures, was performed to characterize the free stream region of turbulent flows through emergent rigid vegetation on rough beds, focusing on turbulence anisotropy. Specifically, the anisotropy invariant maps (AIMs) were determined at different positions within the vegetation array along the flume centerline. The results showed that the combined effect of vegetation and bed roughness causes the evolution of the turbulence from the quasi-three-dimensional isotropy to axisymmetric anisotropy approaching the bed surface. Thus, as the effects of the bed roughness diminish, the turbulence tends towards an isotropic state. Furthermore, it was revealed that also the topographical configuration of the bed surface has a strong impact on the turbulent characteristics of the flow.

2.10. Turbulence in Wall-Wake Flow Downstream of an Isolated Dunal Bedform (Sarkar et al., 2019)

The objective of this study was the analysis of the turbulence in wall-wake flow downstream of an isolated dunal bedform, on the basis of laboratory experiments performed with an ADV. The results revealed that the near-wake flow is featured by sweep events, whereas the far-wake flow is controlled by the ejection events. Downstream of the dune, the turbulent kinetic energy production and dissipation rates, in the near-bed flow zone, are positive. Then, they decrease up to the lower-half of the dune height; beyond that, they increase again. Conversely, in the near-bed flow zone the TKE diffusion and pressure energy diffusion rates are negative; they attain their positive peaks at the crest. Finally, it was found that, below the crest, turbulence has an affinity towards a two-dimensional isotropy, whereas, above the crest, the anisotropy tends to reduce to a quasi-three-dimensional isotropy.

Funding: This research received no external funding.

Conflicts of Interest: The author declares no conflict of interest.

References

1. Chen, K.; Zhang, Y.; Zhong, Q. Wavelet Coherency Structure in Open Channel Flow. *Water* **2019**, *11*, 1664. [CrossRef]
2. Sarkar, S.; Ali, S.Z.; Dey, S. Turbulence in Wall-Wake Flow Downstream of an Isolated Dunal Bedform. *Water* **2019**, *11*, 1975. [CrossRef]
3. Aleixo, R.; Antico, F.; Ricardo, A.M.; Ferreira, R.M. Kinematics of Particles at Entrainment and Disentrainment. *Water* **2020**, *12*, 2110. [CrossRef]
4. Bustamante-Penagos, N.; Niño, Y. Flow–Sediment Turbulent Ejections: Interaction between Surface and Subsurface Flow in Gravel-Bed Contaminated by Fine Sediment. *Water* **2020**, *12*, 1589.
5. Gualtieri, C.; Martone, I.; Filizola Junior, N.P.; Ianniruberto, M. Bedform Morphology in the Area of the Confluence of the Negro and Solimões-Amazon Rivers, Brazil. *Water* **2020**, *12*, 1630. [CrossRef]
6. Penna, N.; Coscarella, F.; D'Ippolito, A.; Gaudio, R. Anisotropy in the Free Stream Region of Turbulent Flows through Emergent Rigid Vegetation on Rough Beds. *Water* **2020**, *12*, 2464. [CrossRef]
7. Lee, S.O.; Hong, S.H. Turbulence Characteristics before and after Scour Upstream of a Scaled-Down Bridge Pier Model. *Water* **2019**, *11*, 1900. [CrossRef]
8. Ben Meftah, M.; De Serio, F.; De Padova, D.; Mossa, M. Hydrodynamic Structure with Scour Hole Downstream of Bed Sills. *Water* **2020**, *12*, 186. [CrossRef]
9. Kang, K.; Lee, G.-H. Man-Induced Discrete Freshwater Discharge and Changes in Flow Structure and Bottom Turbulence in Altered Yeongsan Estuary, Korea. *Water* **2020**, *12*, 1919. [CrossRef]
10. Penna, N.; Coscarella, F.; Gaudio, R. Turbulent Flow Field around Horizontal Cylinders with Scour Hole. *Water* **2020**, *12*, 143. [CrossRef]

Publisher's Note: MDPI stays neutral with regard to jurisdictional claims in published maps and institutional affiliations.

Article

Kinematics of Particles at Entrainment and Disentrainment

Rui Aleixo [1], Federica Antico [1,†], Ana M. Ricardo [1] and Rui M.L. Ferreira [2,*]

[1] CERIS—Civil Engineering Research and Innovation for Sustainability, Av. Rovisco Pais, 1049-003 Lisboa, Portugal; rui.aleixo@tecnico.ulisboa.pt (R.A.); antico@cspfea.net (F.A.); ana.ricardo@tecnico.ulisboa.pt (A.M.R.)
[2] CERIS—Instituto Superior Técnico, Universidade de Lisboa, Av. Rovisco Pais, 1048-001 Lisboa, Portugal
* Correspondence: ruimferreira@tecnico.ulisboa.pt
† Current address: CSPFea Engineering Solutions, via Zuccherificio, 5d, I-35042 Este (Padova), Italy.

Received: 8 May 2020; Accepted: 22 July 2020; Published: 25 July 2020

Abstract: We address the issue of characterizing experimentally entrainment and disentrainment of sediment particles of cohesionless granular beds in turbulent open channel flows. Employing Particle Image Velocimetry, we identify episodes of entrainment and of disentrainment of bed particles by analysing the raw PIV images. We define a reference velocity for entrainment or disentrainment by space-averaging the flow field in the vicinity of the (entrained or disentrainned) particle and by time-averaging that space-average over a short duration encompassing the observed episode. All observations and measurements took place under generalized movement conditions and in non-controlled geometrical set-ups, resulting in unique databases of conditionally sampled turbulent flow kinematics associated with episodes of particle entrainment and of particle disentrainment. Exploring this database, the objective of this paper is to prove further insights on the dynamics of fluid-particle and particle-particle interactions at entrainment and disentrainment and to polemicize the use of a reference velocity to serve as a proxy for hydrodynamics actions responsible for entrainment or disentrainment. In particular, we quantify the reference velocity associated with entrainment and disentrainment episodes and discuss its potential to describe the observed motion *vis-a-vis* local bed micro-topography and the type of entrainment or disentrainment event. Entrainment may occur at a wide range of reference velocities, including smaller than mean (double-averaged) velocities. Anecdotal evidence was collected for some typologies of entrainment: (i) momentum transfer from flow to a single particle, (ii) momentum transfer from a perturbed local flow to a single particle, (iii) collective entrainment associated to momentum transfer between a moving and a resting particle and (iv) collective entrainment considered to be a dislodgment of several particles involving momentum transfer from other particles. In some of these cases, e.g., (ii) and (iii), the use of a reference velocity seems inadequate to characterize the entrainment episode. A word of caution about the use of entrainment models based on reference velocities is henceforth issued and contextualized. In the case of disentrainment, a reference velocity seems to constitute a better descriptor of the observed behaviour. The scatter in the observed values seems to express the contribution of bed micro-topography. All particles were found to experience frictional contacts with the resting bed surface particles, but some particles were stopped more abruptly due to the presence of an obstacle along their path. Most disentrainment of particles took place when the near-bed flow was featuring ejection events.

Keywords: sediment kinematics; entrainment; disentrainment; turbulence

---

1. Introduction

An advanced understanding of bedload transport and fluid-sediment interactions is required to predict how the riverbed evolves in time and what patterns forms in space. Some of the classical formulae for the mean bedload transport are expressed by empirical equations derived by a judicious combination of dimensional analysis, laboratory data production, field data collection and data fitting (e.g., [1–4]). Most were cast within the bounds of theoretical principles, e.g., Bagnold [5], Engelund and Hansen [6], Rijn [7], Wiberg and Smith [8]. The particular path of understanding the physics of individual particles to formulate, employing statistical reasoning, the "laws" that govern the mean or bulk traits of the moving ensemble has been a frequently traveled one, and for which a special mention to Einstein [9,10] is due. It can be argued that Einstein's works constituted the first research program, in the sense of Lakatos [11], addressing, in a way that is still valid today, the physics of mobile sediment boundaries, autonomously from Hydraulics and Geomorphology, and paving the way to modern Fluvial Hydraulics. At the core of his research program is the simple mass conservation statement that the mean bedload transport rate, $\langle q_s \rangle$, can be calculated as the product of the mean entrainment (or pick-up) rate, $\langle E \rangle$ and the mean length traveled by individual particles, $\langle \ell \rangle$, if the variables that describe fluid motion and particle bed mobility and bed morphology are, in a loose sense, in equilibrium, which implies statistical stationarity over a range of time and spatial scales. Symbolically:

$$\langle q_s \rangle = \langle E \rangle \langle \ell \rangle \tag{1}$$

Einstein employed an idealization of turbulent statistics to estimate the probability of the lift force overcoming the particle weight, which was then used to formulate the mean entrainment rate. It is in this limited sense that Einstein's (1950) bedload formula is probabilistic, while having far reaching impact over subsequent research. Engelund and Fredsoe [12], Cheng and Chiew [13], among others, refined the model for the probability of lift exceeding particle weight. Other authors proposed more complexity in the description of the geometry of the destabilized particle and corresponding hydrodynamic actions [8,14,15]. In this line of thought, Ferreira et al. [16] attempted to articulate Einstein's Equation (1) with Paintal [17] probabilistic view that the mean bedload transport rate is determined by the probability density functions (pdfs) of the bed shear stress (representing the destabilizing effect) and of the resisting forces per unit bed area. The latter probability density expresses grain resistance as it varies with bed structure (Schmeeckle et al. [18]—resistance not futile). The pdf of flow velocities is then not considered universal but a function of the particular bed structure. The marginal probabilities of exceeding the threshold velocity for entrainment are integrated for all parameterized bed conditions. The bedload formula of Ferreira et al. [16] assumes that entrainment occurs when the destabilizing hydrodynamic actions overcome the stabilizing forces originated by particle weight and local geometry. This is incomplete, at best, as it was demonstrated beyond doubt that entrainment requires that the particle overcomes a potential energy wall, thus leaving its pocket to move unconstrained [19]. In other words, the threshold set by force or momentum equilibrium is a necessary but not sufficient condition for entrainment, a finite period of time is necessary to transfer enough momentum to the particle for actual entrainment to take place [20].

Defining the threshold for entrainment based on the work of the vertical forces to overcome the potential energy wall does not solve the fundamental problem associated with the project of deriving mean bedload transport formulas based on the probability of exceeding that threshold. The most serious caveat is very limited knowledge of the statistical representation of the hydrodynamic actions on bed particles, despite the early efforts by Chepil [21] and recent CFD-based studies (e.g., [22]). To compensate for this lack of knowledge, fluid flow velocities in the vicinity of the particle—a competent velocity, to use a classical term [23]—were used as proxies of forces, generally converted into forces through the use of lift and drag coefficients [14,16,19,24,25]. This approach overlooks the obvious difficulty that fluid momentum is passed on to the particle, in a finite time interval, through the development and adjustment of the particle boundary layer and eventual lee flow separation, generating viscous stresses and pressure imbalances that can be integrated into forces on an appropriate coordinate system. This process involves mobilizing fluid inertia, associated with boundary layer "memory", and may have different outcomes depending on the shape of the particle [25], its geometrical arrangement and local bed micro-topography [18]. Hence, the same flow may or may not be able to entrain a sediment particle. A model based on the concept of competent velocity would capture the correct outcome only if it could estimate the correct values of the drag and lift coefficients specific for that particle and location.

This is a major concern and one of the key motivations of this study. Formulas developed from sound grain-scale physical principles and well-formulated probabilistic techniques have not been shown to perform significantly better than more *ad hoc* empirical approaches presumably because they also rely on parameters for which information is insufficient. In general, whether purely empirical or physically based, all mean bedload transport rate formulas may fail to predict the actual bedload rate by one order of magnitude [26], when tested with data outside their calibration range. We believe that the mean entrainment rate $\langle E \rangle$ can be a key ingredient to construct better physically-based formulas within a probabilistic paradigm but only if the interaction of fluid flow and bed particles is better understood, which calls for a closer observation.

We also note that the mean bedload transport rate is not enough to characterize morphology and sediment dynamics of a stream, as bedload fluctuations may be more than 10-fold the mean bedload discharge rate [27]. To make matters still more complex, the time or space windows employed to, in practical terms and avoiding ergodicity issues [28], define the mean bedload transport rate may introduce bias in its value and certainly determine the quantification of the fluctuations.

Fluctuations in the value of the bedload transport rate arise from imbalance between entrainment and disentrainmemt rates. They are not necessarily caused be turbulence, as they were registered in laminar fluid flows [29], and are probably associated with positive feedback effects born out of particle-particle interactions. Ancey et al. [30] was able to retrieve this highly fluctuating behaviour (actually, non-Gaussian) by employing a birth–death–emigration–immigration Markov processes to estimate the probability of registering a given number of particles moving in a finite control volume at a given instant. In this model, the entrainment rate includes two processes: momentum transferred directly from the fluid flow or momentum imparted by moving particles. In this context, collective entrainment came to signify the process whose rate was proportional to the number of particles already in motion.

Positive feedback (increased entrainment due to collective motion) and also negative feedback (particle disentrainment due to interactions with bed), both promoting the enhancement of the magnitude of bedload fluctuations, were seen to cause clustering [31] and induce the onset of bed instability leading to the formation of sediment waves, if moving patches created by the local imbalance between entrainment and disentrainment grows beyond a critical height [32]. The motion of sediment waves and, in general, the evolution of bed morphology, for instance as a response to flow unsteadiness or in gradually varied flows, is described by Exner equation, which in "entrainment form" can be written as:

$$(1 - \lambda) \frac{\partial Z_b}{\partial t} = -(E - D) \tag{2}$$

where Z_b is the local bed elevation (loosely, the elevation of the crests of the non-moving particles), and D and E are the disentrainment and the entrainment rates, respectively, and λ is the bed surface void fraction. The conservation of the mass of moving particles is linked to Equation (2) as:

$$\frac{\partial h_s}{\partial t} + \frac{\partial h_s u_p}{\partial x} = E - D \tag{3}$$

where h_s is the particle activity (volume of particles in motion per unit streambed area, [24,33]) and u_p is the particle velocity.

Realistic bed forms, including longitudinal bars and dunes or anti-dunes, can be generated computationally within the numerical solution of morphological models based on the shallow-water equations and on Equations (2) and (3) but not for all formulations of wall resistance and bedload transport rates [34]. Of special significance for this text is the fact that bed forms can be generated if the actual bedload discharge rate at a given instant and location is not an injective function of the flow velocity at that instant and location [35]. This can be achieved by expressing the imbalance between E and D in Equation (2) as relaxation source term in the form $\alpha\left(q_s - q_s^*\right)$ where q_s^* is an "equilibrium" or "saturation" bedload discharge (as opposed to the actual bedload discharge q_s) and α is a relaxation parameter [36–39]. The issue of "non-locality" of bedload transport was formulated within a probabilistic framework by Furbish et al. [33], Bohorquez and Ancey [40], among others. Starting from a differential account of the Markovian birth–death–emigration–immigration processes or from a definition of local bedload discharge rate as $q_s = h_s u_p$ in which each factor can be decomposed into an average and a fluctuating part, the main result is that the mean bedload transport rate can be expressed as:

$$\langle q_s \rangle = \langle h_s \rangle \langle u_p \rangle + \frac{\partial}{\partial t}\left(\varepsilon h_s\right) \tag{4}$$

where ε is a particle diffusivity. The second term in the left hand side of Equation (4) is a diffusive bedload contribution and guaranties "non-locality", i.e., that the mean bedload discharge is not a simple function of the ensemble averaged local flow velocity.

To emphasise this point of non-locality, Furbish et al. [28] argue that admitting a fluctuating component of particle velocity amounts to admitting that only non-local transport exists, i.e., "particles moving across a surface at any instant in time (...) started their motions 'nonlocally' from many positions and previous instances". While it is certainly true that the morphological consequences of the imbalance between entrainment and disentrainment are seen only down the stream, there is a fundamental issue of "locality" to be addressed—the fact that one needs to close the source term $E - D$, which should involve (or may benefit from involving) considerations on the local fluid flow field and particle motion and how momentum is imparted to particles resulting in entrainment. This configures the second main concern that motivates this research—the condition of possibility to express the imbalance between E and D based on the local flow field and, in particular, based on a competent velocity.

In this respect, we note that entrainment was subjected to a great deal of attention while disentrainment was the object of very little dedicated research. To the best of our knowledge, the most complete experimental description of disentrainment processes can be found in Cecchetto et al. [41]. They investigated experimentally the role of both the flow field and the bed arrangement in the disentrainment of bedload particles and found that lower values of instantaneous longitudinal velocities were linked to disentrainment events. Investigating the disposition of resting sediments over an area, they found that the depositional processes are driven by bed roughness and that deposited grains were characterized by a non-random spatial distribution. Given that the measurements of Cecchetto et al. [41] were taken at more than 1.5 particle diameters above the bed, it may be difficult to formulate a link between local flow velocity and disentrainment. Yet, while one can develop a formula for the mean bedload transport rate based on the assumption $E = D$ [10], to deal with unsteady flows or complex stream morphologies, it is necessary to close both E and D in Equation (2).

Acknowledging this state of affairs, we propose a step back to observe actual entrainment and disentrainment events of sediment particles in turbulent open-channel flows over cohesionless granular beds. We believe this observation constitutes a preliminary step in the formulation of a theory that may or may not involve a reference velocity to express the deterministic threshold of entrainment or disentrainment. We propose to conduct this observation in a transport system purged of many accessory complexity (at this stage)—a "minimal system", in the sense of Ancey [42], that features spherical smooth particles arranged in a lattice, while at rest in the bed, in a prismatic channel that is wide enough to render negligible the effects of lateral walls and secondary currents. The observation protocol was designed in order to be the least intrusive as possible. We do not place particles in controlled exposed positions. Instead, we define an observation window, we observe the flow as imaged by a laser sheet and record its configuration every time an entrainment or disentrainment event occurs. We thus seek to collect as much information as possible from events that spontaneously take place in the observation window. In the case of entrainment, we obtained databases in generalized transport conditions which is relatively rare, compared to those obtained under incipient motion conditions (in the sense of Kramer [43]) or obtained with a test particle in an otherwise fixed bed. Our databases possess the advantage of allowing for the discussion of the interactions among moving particles. Furthermore, for the objective of studying disentrainment, our observation protocol is particularly adequate. In this sense, for both entrainment and disentrainment, the observations allow for a unique discussion of the merits and difficulties of employing a reference velocity as a proxy for hydrodynamic force or power.

We thus aim at providing further insights on the dynamics of fluid-particle and particle-particle interactions at entrainment and disentrainment and to polemicize the use of a reference velocity to be used as a proxy for hydrodynamic actions responsible for entrainment or disentrainment. It must be stated that the goal is not to infer a general model of particle entrainment/disentrainment, but rather to have a more complete picture of the fluid-particle and particle-particle interactions naturally occurring in the flume's bed.

The paper is organized as follows: Section 2 is dedicated to the description of the experimental setup and procedures. In Section 3 the concept of reference velocity is discussed and two approaches to define it are presented. In Section 4 the obtained experimental results are presented with focus on (i) the evaluation of the critical assessment of a reference velocity as a proxy of hydrodynamic actions to describe entrainment events, on (ii) the detailed discussion of 4 different types of entrainment events and on (iii) the reporting and analysing of disentrainment events. The paper is closed by a set of main conclusions, Section 5.

2. Experimental Setup

The experimental work was carried out in a 12.5 m long and 0.408 m wide glass-sided flume at the Laboratory of Hydraulics and Environment of Instituto Superior Técnico, Lisbon. The initial 7 m long fixed-bed reach comprised 1.5 m of large boulders (50 mm average diameter), 3.0 m of smooth bottom (PVC) and 2.5 m of one layer of glued spherical glass beads (5.0 mm diameter); 4 m of the remaining flume were filled with 5 layers of 5.0 mm diameter glass beads, with density $\rho_s = 2490\,\text{kg}\,\text{m}^{-3}$, packed (with some vibration) to a void fraction of 0.356, typical of random packing. Figure 1 depicts the packed loose bed.

Figure 1. Example of the loose bed glass spherical particles with 5 mm diameter.

Water and sediments were recirculated through independent circuits. At the flume outlet sediment particles are collected and recirculated through a dedicated pumping circuit. This dedicated circuit transported the sediment particles from the flume outlet back to the flume upstream, dropping the sediments through the free surface at the section $x = 3$ m measured from the water inlet. To measure the solid flow rate, a particle counter device capable of measuring bedload discharges was installed at the downstream end of the mobile bed reach. Details on the particle counter device (mechanical system and firmware), included device's installation and validation, can be consulted in Mendes et al. [44].

A 2-D component Particle Image Velocimetry (PIV) system was employed both for a general characterization of the flow field of the experimental tests (obtaining longitudinal u and vertical w instantaneous flow velocities) and for the spatial and temporal definition of the flow velocities associated with entrainment and disentrainment events. In the former case, the observation window covered the entire flow depth and a length comprises between 6 cm and 12 cm, depending on the test. The duration of each observation was 5 min corresponding to 4500 image couples. The measurements were carried on the channel centerline, located at 20.4 cm from the channel window. An extra test PIV acquisition run with a duration of 20 min was carried out to compute statistics of possible large scale velocity fluctuations.

The PIV system consisted of an 8 bit $1600 \times 1200 \, \text{px}^2$ CCD camera and a double-cavity Nd-YAG laser with pulse energy of 30 mJ at wavelength of 532 nm. The system was operated at 15 Hz with a time between pulses within the range from 380 µs to 500 µs. Polyurethane particles with mean diameter of 50 µm in a range from 30 µm to 70 µm and specific density of $1.31 \, \text{g cm}^{-3}$ were used as seeding. Dantec Dynamics' DynamicStudio software allowed for processing image pairs with adaptive correlation algorithm. The initial interrogation area was of $128 \times 128 \, \text{px}^2$, while the final was of $16 \times 16 \, \text{px}^2$, with an overlap of 50%. Corresponding to a spatial resolution of 0.26 mm for a field of view (FoV) of $6 \times 6 \, \text{cm}^2$, and a resolution of about 1 mm for the wider field of view tested ($6 \times 12 \, \text{cm}^2$). An acetate sheet was placed on the water surface to ensure optical stability and absence of laser sheet reflections. The presented experiments corresponded to the case of a nearly uniform flow. Free surface elevation and bed level were measured with 0.5 mm resolution point gage in 5 transversal sections of the flume and in 3 lateral positions per cross-section. Two experimental tests, T1 and T2, whose flow characteristics are reported in Table 1, are presented in this paper. Variables in Table 1 are the flow discharge, Q, the mean flow depth, h, the depth-averaged mean longitudinal velocity, U, the friction velocity and bed shear stress calculated from the vertical turbulent stresses profile, u_* and τ_b, respectively. Non-dimensional hydraulic parameters are the Froude number, $Fr = U/\sqrt{gh}$, Shields parameter, $\theta = u_*^2/(s-1)gd$, Reynolds number of the mean flow, $Re = Uh/\nu^{(w)}$, bed Reynolds number, $Re_* = ud/\nu^{(w)}$ and the non-dimensional mean bedload discharge, $\Phi = \langle q_s \rangle / \sqrt{(s-1)gd^3}$, where d is the particle diameter, g is the gravitational acceleration, $\nu^{(w)}$ is the water kinematic viscosity and $s = \rho_s/\rho_w$ is the ratio between the sediment and water densities. As seen in Table 1, the experiments are conducted in generalized transport conditions.

Table 1. Hydraulic parameters characterizing the mean flow.

Test	Q ($\text{m}^3\,\text{s}^{-1}$)	h (m)	U (m s^{-1})	u_* (m s^{-1})	τ_b (Pa)	Fr (-)	Re (-)	Re_* (-)	θ (-)	Φ (-)
T1	0.01667	0.0684	0.6016	0.0475	2.2557	0.7345	4.61×10^4	267.63	0.0301	0.0007
T2	0.02135	0.0696	0.7574	0.0610	3.7250	0.9166	5.90×10^4	337.06	0.0497	0.0034

The mean bedload rate was determined from the time series of particle hits which for each test, comprised more than 10 consecutive hours of observations.

Entrainment and disentrainment events were not imposed. The present approach consisted of detecting and identifying, by means of a careful visual inspection of the PIV images, individual sediment particle at entrainment and at disentrainment events, naturally occurring in the PIV field of view. The dynamic conditions of these naturally occurring events were then measured. A total count of 44 particle dislodgments (15 in test T1 and 29 in test T2) and 11 particle disentrainments (test T1). From these set of data, 6 events were chosen as example and presented in the next section.

3. The Reference Velocity

The proposed observation protocol comprised the monitoring of the turbulent flow field in proximity of the crest of the particles about to be entrained or just disentrained. There were no test particles placed or pre-arranged geometries. Observations were meant to acquire information from sediment particles that spontaneously were entrained or disentrained while located on the plane of the PIV laser sheet. A visual inspection of the PIV footage allowed to include in the analysis only particles entrained resulting in brighter shade and discard those not properly illuminated by the laser sheet.

A competent velocity for entrainment is often the key element that allows for a practical formulation of the energy balance [19] or the limit force or momentum equilibrium for one particle at destabilization conditions (e.g., [45]). This velocity is normally measured in the vicinity of the particle susceptible to be entrained. We call this a "reference flow velocity", sampled from the particle near-field.

We decided that the reference velocity should be susceptible to be converted into a drag or lift force, through the application of suitable coefficients (e.g., [46]), and susceptible to be measured with no special or intrusive apparatus. Considering different possibilities [47], and considering we did not want to employ test particles, we opted to measure above the specific particle that we had seen being entrained or disentrained. We tested but we ruled out measuring in front of the particle, namely at the elevation of the plane of its equator, because the view to that position was frequently obstructed by other particles out of the laser plane located above the particle crest at a certain reference height.

We thus opted to measure at approximately an elevation $d/2$ above the crest of the entrained particle (defined in the last frame where it appears immobile) or of the disentrained particle (defined at the first frame where it appeared immobile), as seen in Figure 2. This elevation can be considered to be a compromise between the quality of the data and the adequateness to represent the near-particle flow. Defining the reference velocity nearer the crest of the particle would make it susceptible to bad data, as seeding depletion, lower velocities and reflections from bed particles affect negatively the PIV signal. Measuring further above might reduce its explanatory value as the shear rate of the double-average longitudinal velocity is very high in this near-bed region. As an example, had we measured at 8 mm above the crests, as in Cecchetto et al. [41], the double averaged velocity would be 37% larger.

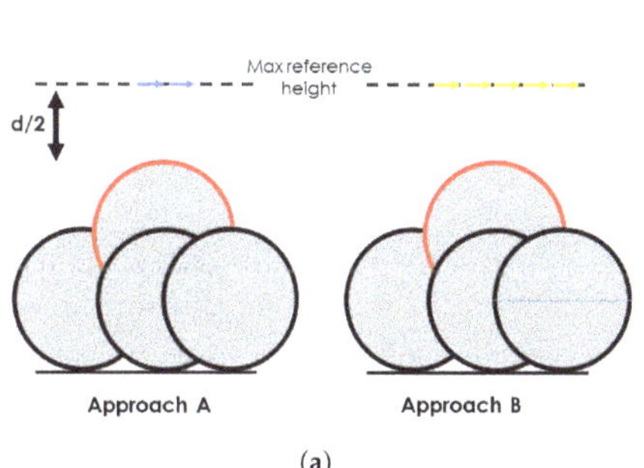

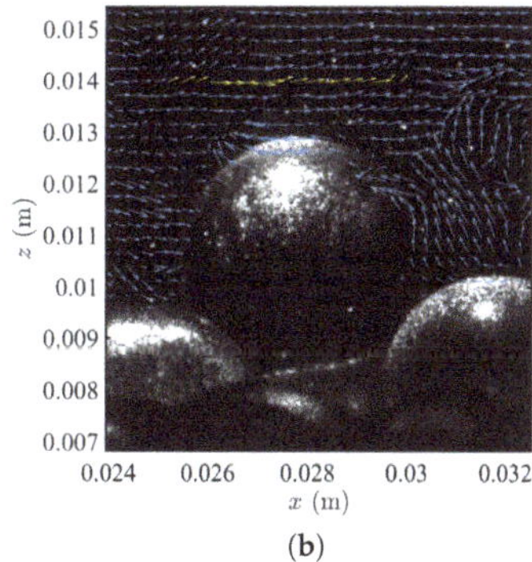

Figure 2. (**a**) Scheme of the velocity vectors considered in the spatial average in approach *A* and *B*; velocity vectors averaged in approach *A* are reported in blue, while those averaged in approach *B* in yellow. (**b**) PIV image with velocity vectors superimposed; velocity vectors averaged in approach *B* are reported in yellow.

We underline that the reference elevation was not fixed. We stand by this option as it is makes the reference velocity comparable among entrainment events and it is not difficult to enforce, either in post-processing PIV data or in tests with pre-arranged geometries. We note that we opted to use the

PIV data closer to the elevation $d/2$ above the crest of the particle (but always below that elevation) instead of interpolating the data at exactly the plane $d/2$. This again is a compromise imposed by the (sometimes poor) quality of the data at lower interrogation area rows (Figure 2). Finally, we opted to consider that the reference velocity is a time average of the instantaneous velocities acquired across the entrainment or disentrainment event. This average involves the last observation in which the particle was immobile/moving and the first two for which it was moving/immobile, respectively for entrainment/disentrainment. Given the PIV time rate (15 Hz) this means that the reference time window is $2/15 \approx 0.133$ s.

Reference flow velocities at the specified reference elevation were computed by two different methods:

1. Approach A: as the spatial average between the two velocity vectors in adjacent interrogation areas located above the top-center of the particle.
2. Approach B: as the spatial average of velocity vectors of all interrogation windows directly above the particle (spanning its diameter) located at the same reference height of approach *A*.

A scheme of the two approaches is depicted in Figure 2a, while Figure 2b depicts a PIV image with velocity vectors superimposed: instantaneous velocity vectors are represented in blue, whereas those averaged in approach *B* are reported in yellow.

The reference velocity vector has two orthogonal components in the wall-normal and in the along-wall directions. We use the later to serve as proxy for hydrodynamic forces. However, it may be relevant to know what kind of contribution to the shear stress is associated with the measured reference velocity and it is surely relevant to understand specific events whether the motion is characterized by velocities higher or lower than the double-averaged velocity above the plane of the crests. For that purpose, we employed a quadrant analysis to jointly discuss u' and w', the fluctuations of the along-wall and wall-normal components of the reference velocity. We employ the usual terms Nakagawa and Nezu [48] of outward interaction ($Q1$, $u' > 0$, $w' > 0$), ejection ($Q2$, $u' < 0$, $w' > 0$), inward interaction ($Q3$, $u' < 0$, $w' < 0$) and sweep ($Q4$, $u' > 0$, $w' < 0$). Please note that since the reference velocity is a space-time average at the scale of the particle, the contributions to the space-averaged instantaneous shear stress are only approximate.

As for the double-averaged velocity, we consider the intrinsic space-time average of the flow field at an elevation equal to d/s above the initial spatially averaged bed particles crest level, as seen in the PIV calibration images. The intrinsic average was obtained from the superficial average by dividing by the the space-time porosity $\phi_{VT}(x_i, t)$, representing the ratio of fluid to total averaging domain [49]:

$$\phi_{V\,T}(x_i, t) = \frac{1}{T_0} \frac{1}{V_0} \int_{T_0} \int_{V_0} \gamma\,(x_i + \xi_i,\, t + \tau)\,dV d\tau \tag{5}$$

where T_0 and V_0 are respectively the averaging period and the averaging domain, and $\gamma = 1$ if the region (x_i, t) is occupied with fluid and $\gamma = 0$, otherwise. Please note that the reference velocities for entrainment and disentrainment are not acquired necessarily at the elevation of the double-averaged velocity, as it depends on the elevation of initial/resting position of the crest of the entrained/disentrained particle. The difference, however, is small and since the bed surface did not develop bedforms and remained essentially planar.

4. Observations

4.1. The Big Picture

To characterize possible turbulent large scales responsible for relevant fluctuations of particle activity, we measured flow velocities for 20 min in the central plane of the channel. Focusing on near-bed processes, we averaged the velocities registered at an elevation of $d/2$ above the crests of the initial bed and we averaged in space over the length of the PIV FoV (about 10 cm). We further averaged these velocities on intervals of 1 s, for reasons that will be clear next. A times series of this space-time filtered velocity is shown in Figure 3.

The bedload discharge meter described in Mendes et al. [44] was employed to measure a long time series of bedload discharge rates. As it is a particle counter, we opted to group counts registered in a 1 s interval. This justifies the choice of the averaging window employed on the velocity time series. The resulting time series of bedload discharge can be observed in Figure 4. The time average applied to these measurements work as a low-pass filter, keeping only large-scale fluctuations. The additional spatial average applied to the velocity measurements was meant to compensate for particle diffusion between the location of the measurements and the location of the pressurized boxes where bedload was measured. A moving-average filter with a window of 21 s was then employed and its results superimposed to the previous time series, just to guide the eye and underline the main fluctuations in Figures 3 and 4.

Observing Figure 3 it is evident that this low-pass filtered velocity exhibits periods where it consistently remains above the local average for several seconds and also below the average for several seconds. We do not know the origin of these very large fluctuations (we do not know, in particular, if they are very large-scale of motion (VLSM), in the sense of Kim and Adrian [50]).

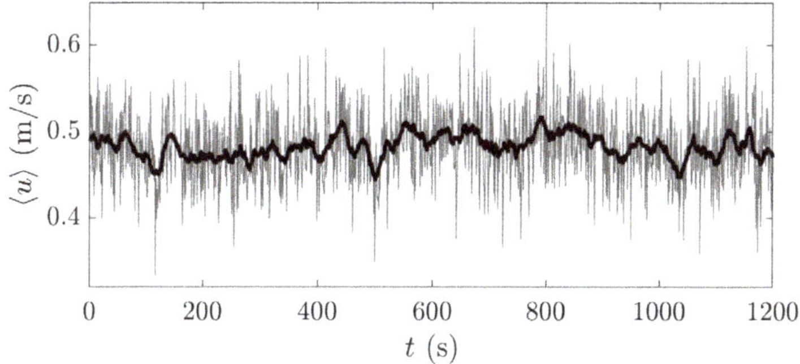

Figure 3. Velocity time seriesfor test T2 at $d/2$ above the crests of the initial bed averaged in space over the length of the PIV FoV.

Given its time permanence, we did expect to see an increase or a decrease in particle activity (and bedload transport rates) associated with these fluctuations. Observing Figure 4, we rest assured that these large scale fluctuations are also present in the bedload time series. These measurements were not synchronous, so we could not measure the time correlation, but a simple inspection reveals that both series exhibit sustained periods of lower than average and higher than average values, connected by equally long, albeit intermittent, transitions.

Therefore, the overall picture is that there is a general agreement between flow momentum and particle activity. That is fully in agreement with empirical formulas that were derived from this principle [51].

A closer look reveals that the fluctuations of the bedload discharge are much larger than those of the filtered velocity. This, again, is in accordance with what is expected from the bulk behaviour of water streams, as the bedload transport rate is a non-linear function of the of fluid flow variables. For instance if the particle rate was formulated as a function of the third power of the flow velocity, the observed maximum fluctuations of 35% on the flow velocity would be translated into 140% fluctuations of bedload, which matches reasonably well the observed maximum fluctuations of the latter.

However, analysing the probability distribution functions (pdf) of these low-pass filtered signals—Figure 5 for the case of velocities and Figure 6 for the case of the bedload transport rate in the central 10 cm, Q_s,—one finds a fundamental difference. The velocity pdf is symmetrical, while the pdf of the bedload is (slightly) positively skewed with a longer tail events much larger than the mean.

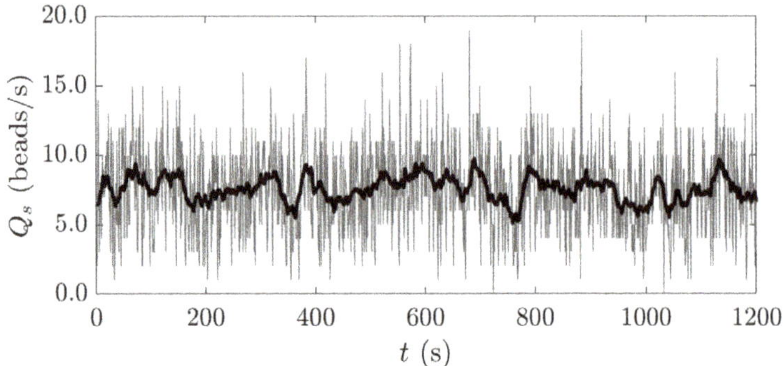

Figure 4. Sediment discharge time series measured during test T2 at the channel centerline with the bedload discharge meter described in Mendes et al. [44].

This can be a result of the positive feedback arising from collective entrainment [30]. Higher near-bed velocities induce a stronger particle activity which, in turn, induces even larger activity by direct momentum transfer among particles, amplified local turbulence or increased exposure. Perhaps the negative feedback envisaged by Cecchetto et al. [31], as particles seem to be stopped where other particles the have stopped before, forming clusters, is not so effective.

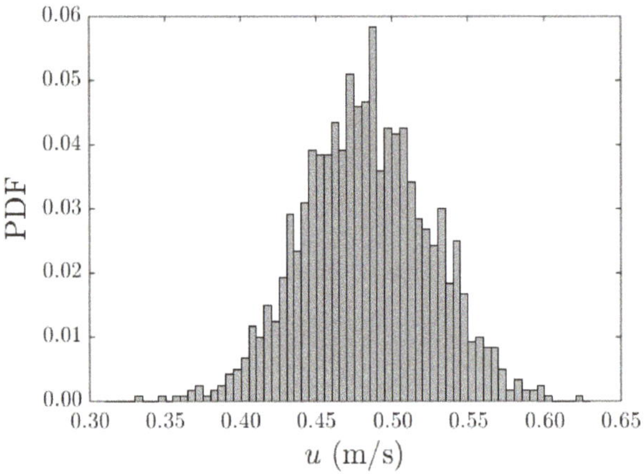

Figure 5. Normalized histogram of the lowpass filtered longitudinal velocity obtained at the reference elevation in test T2.

In any case, should positive feedback be relevant, it may constitute a difficulty for models that employ a competent velocity as a proxy for force or energy thresholds of entrainment. It may be the case that, when particle activity is larger, a relevant proportion of moving particles had been entrained by processes other than momentum transfer from the fluid.

We note, however, that there is no strong empirical evidence that allows for a quantification of the modes of collective entrainment. Simultaneous entrainment of several particles, as a result of sweep events, has been reported in the literature for several decades now (e.g., [52–54]). Enhanced motion due to particle-particle interactions has also been well documented [30,55], but an empirical formulation of the rate of collective entrainmemt, as a linear function of particle activity or otherwise, has not been achieved yet. This calls for a closer observation of particle entrainment and, given that bedload fluctuations are caused by the imbalance between entrainment and disentrainment rates, a direct observation of disentrainment becomes also necessary. We offer a contribution, along these lines, in the next section.

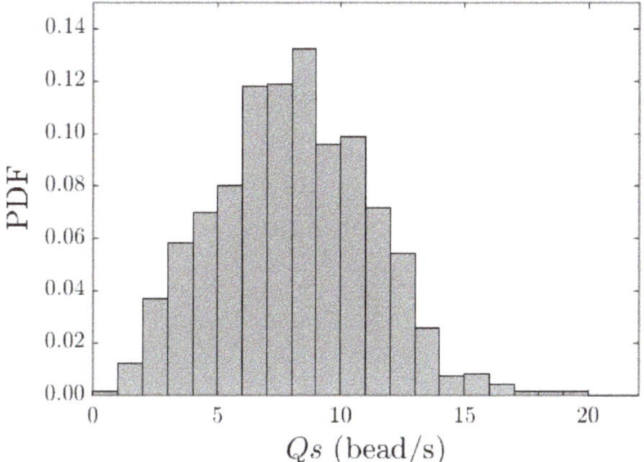

Figure 6. Normalized histogram of the bedload discharge, Q_s, in particles per second, registered, in test T2, across the central 10 cm of the flume.

4.2. Local Observations of Entrainment

The first question to be answered with our observations concerns the magnitude of the competent velocity for entrainment, namely if entrainment is associated with exceptional events only or it may occur at relatively mild velocities. Figures 7 and 8 depicts the reference velocity fluctuations u' and w' associated with particle entrainment event (red dots) for tests T1 and T2. Figure 7a,b are relative to approaches A and B, respectively, applied to test T1, and Figure 8a,b are the same, applied to test T2. In the background of these quadrant plots are the contour lines of the two-dimensional histogram of temporal fluctuations of the spatially averaged velocity around the double-averaged velocity at the initial reference height. To compute the double-averaged velocity, 2,729,484 data points were considered in test T1 and 3,917,061 in test T2.

Entrainment occurs at a wide range of reference flow velocities, but a prevalence of sweep and outward interactions (longitudinal velocities larger than the mean) can be observed for both tests. Some particles were entrained even in the presence of negative values of longitudinal velocity, i.e., they were entrained by flows with velocities lower than the mean flow. However, in this case, fluctuations associated with negative vertical fluctuations w' are rare. Most of the recorded lower-than-mean flow entrainments were associated with ejections.

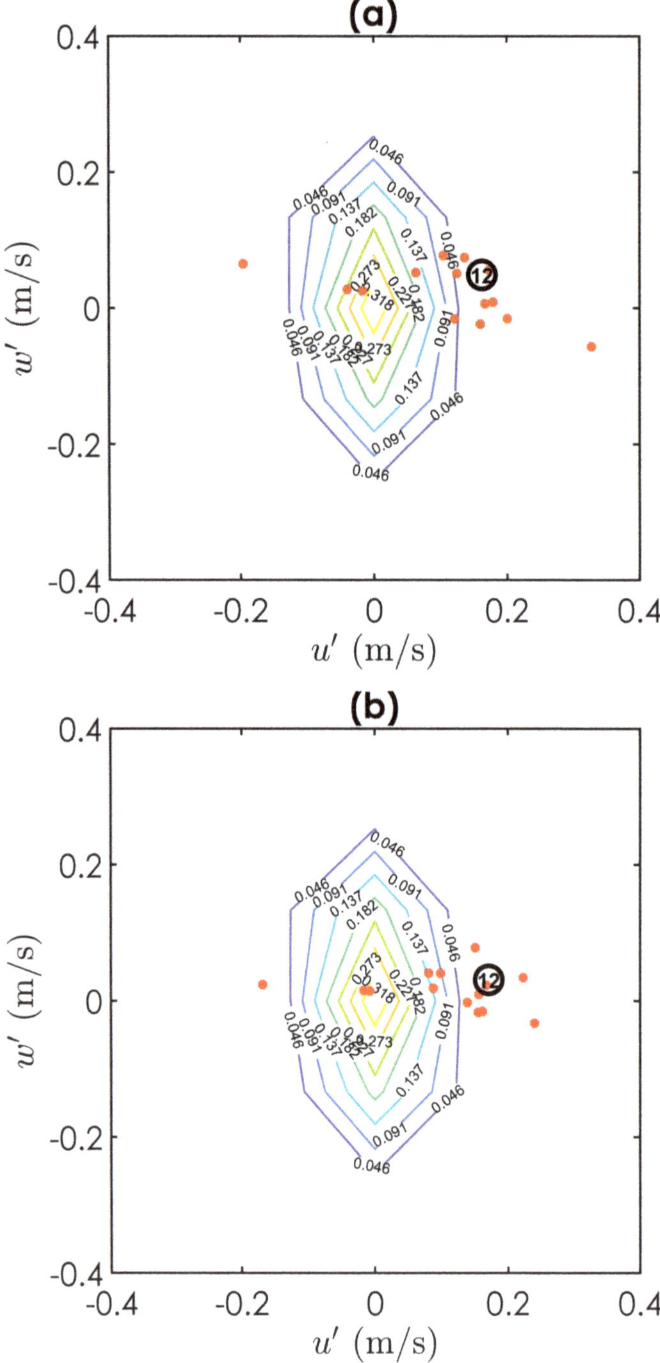

Figure 7. Reference velocity fluctuations (red dots) characterizing: (**a**) test T1-Approach A; (**b**) test T1-Approach B. Contour lines represent the 2D histogram of the velocity time fluctuations at the reference height for the entire set of PIV images.

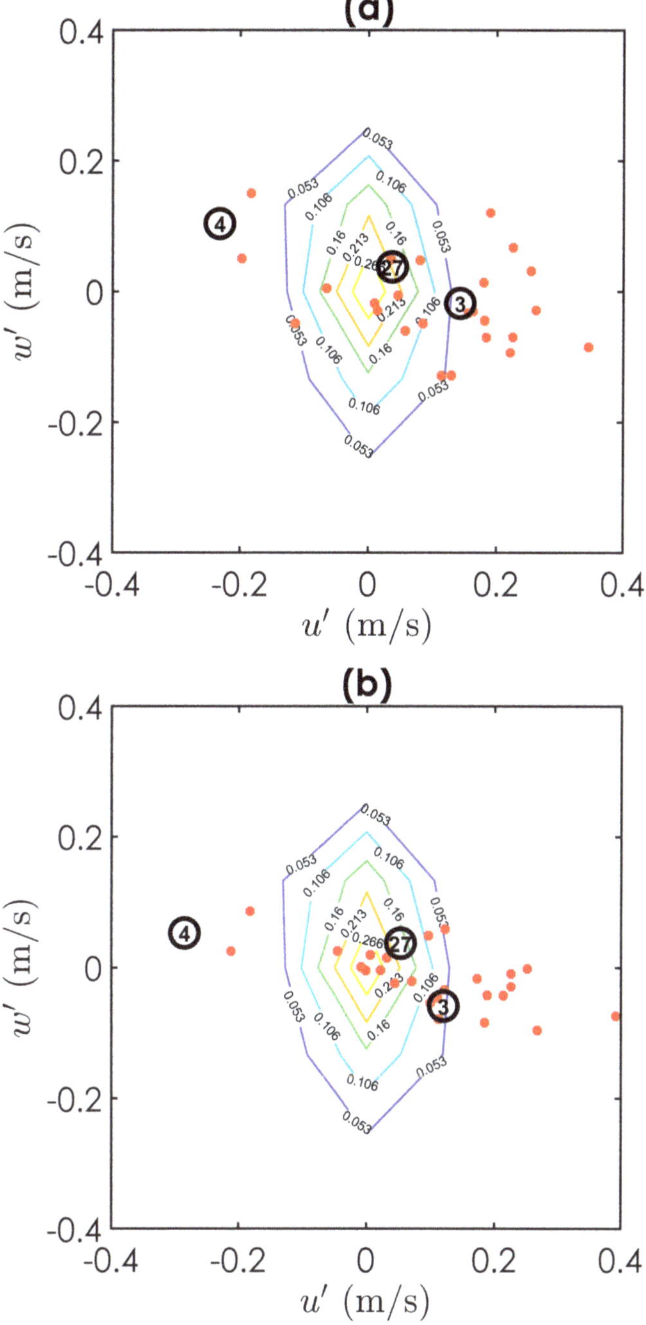

Figure 8. Reference velocity fluctuations (red dots) characterizing: (**a**) test T2-Approach A; (**b**) test T2-Approach B. Contour lines represent the 2D histogram of the velocity time fluctuations at the reference height for the entire set of PIV images.

For identical particles with the same skin roughness, geometry of contacts in the lee side and pivoting axis, as illustrated in Cases 1 and 2 of Figure 9, the forces promoting their destabilization are the same [16]. In that case, differences in the competent velocity for entrainment can only be due to different drag and lift coefficients, due to different boundary layer development histories, different exposures (Cases 1, 2 and 4 of Figure 9) or different persistence of the hydrodynamic actions. In this latter case, Diplas et al. [20], Valyrakis et al. [56], among others, argued that the presence of peak values in the fluctuating hydrodynamic actions is not a sufficient condition for particle entrainment: grain removal is closely related with the duration of energetic near-bed turbulent events, namely the impulse of hydrodynamic forces (the momentum variation). Exceeding a certain critical magnitude value is not enough to promote entrainment. Sufficient momentum transfer is needed to remove the particle out of its initial bed location, or as Valyrakis et al. [19] formulates it, the work of the hydrodynamic forces has to be sufficiently high to endow the particle with enough potential energy to place it above the crest of the downstream obstacle. In this case, an adequate parameterization of the local bed micro-topography, including the geometry (and the spatial probability distribution) of the downstream obstacles would still allow for using a competent velocity as a proxy of a threshold of hydrodynamic actions. Particle exposure, here defined as the difference between the height of the crest of the neighbor particle located upstream and the crest of the particle about to be entrained (Figure 9), is evidently a major influence for particle entrainment. The same near-field fluid flow may be effective or ineffective for entrainment depending on the exposed area available to effective hydrodynamic actions. The problem, however, can be more complex, since particles with the same exposure (Figure 9 cases 1 and 3) may develop quite different pressure differences between the front and lee sides, and hence different pressure drag contributions, depending on the influence of the geometry of the downstream neighbors on boundary layer separation. A major difference between Cases 1 and 3 of Figure 9 is evidently the downstream support plane and pivoting axis but that can be taken into account by an adequate formulation of the geometry in the force (or their work) or momentum balance at threshold conditions. The variation of the lift or drag coefficient is a more complex matter for which available information is still scarce. In this respect, the work of Dwivedi et al. [46] is a major step forward. Should these advances on the parameterization of lift and drag coefficients be able to be integrated in threshold models, the case for the use of a competent velocity for entrainment is strengthened.

Yet more subtle is the effect of the history of boundary layer development on individual particles. It may be the case that identical near-bed flows acting on identically exposed particles produce different hydrodynamic actions, depending on the past of the flow and local geometry. For instance, a particle might find itself exposed, e.g., because of the removal of an upstream neighbor, inducing a very large pressure imbalance and, ultimately, entrained. A particle exposed similarly and experiencing a gradual build-up of fluid velocity might not experience such a strong pressure imbalance, as the lee side flow might have time to adjust, and hence the particle would remain in the bed. This is merely speculative, we do not have data to quantify this possibility, but we believe it is an argument to keep in mind when discussing the possibility of finding an adequate reference velocity to serve as competent velocity for entrainment.

Should our choice of reference velocity be adequate to express the hydrodynamic actions registered at entrainment, the exposure observed at the instant of entrainment should match the exposure calculated by a theoretical model with that reference velocity as competent velocity. A visual inspection of the PIV images enabled computing the particle exposure as the difference between the entrained particle crest level before entrainment and the crest level of the closer particle located immediately upstream for all the sample of entrainment events. The model by Ferreira et al. [16] was employed to determine the theoretical exposure, assuming that the measured reference velocity is the theoretical competent velocity for entrainment u_p, i.e., the velocity that expresses a threshold condition for particle stability, given local geometry and drag and lift coefficients. For spherical particles in nearly horizontal streams, the model equations are

$$C_e = \sqrt{\frac{4}{3}\frac{(s-1)\,gd}{u_p^2}\frac{\tan(\varphi)+\tan(\psi)}{(1-\tan(\varphi)+\tan(\psi))+\frac{C_L}{C_D}(\tan(\varphi)+\tan(\psi))}}, \tag{6}$$

for a force threshold (sliding instability), or

$$C_e = \sqrt{\frac{(s-1)\,gd}{u_p^2}\frac{\tan(\theta)}{1-\tan(\theta)}}, \tag{7}$$

for a moment threshold (rolling instability). Variables and parameters in Equations (6) and (7) are C_e, the exposure coefficient (defined above), $(s-1)$, the submerged particle specific gravity, g the acceleration of gravity, d, the particle diameter, C_L and C_D, lift and drag coefficients, respectively, φ, the particle's angle of support, ψ the angle expressing skin roughness and θ, the angle between the vertical plane and the plane that encompasses the centre of mass of the particle and the pivoting axis.

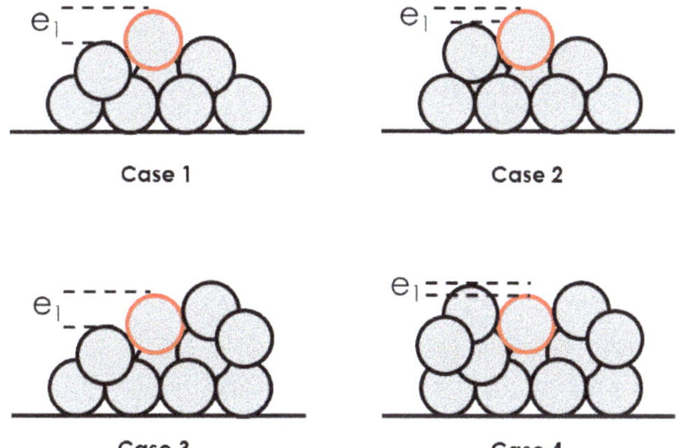

Figure 9. Influence of particle bed topography on particle entrainment. Four cases are reported here, in which the particle at entrainment is identified by red contours and e_1 represents the particle exposure: in case (1) the particle at entrainment is characterized by greater exposure with respect to case (2); in case (3) the exposure is the same as in case (1), but the presence of an obstacle downstream hampers particle entrainment—for the same particle-exposure and fluid pressure a larger impulse is needed to overcome the potential energy wall created by the protruding downstream particle. Finally, at case (4), the particle has a negative exposure and a protruding downstream particle; it will require a strong lift force, sustained in time so that its work is able to overcome the potential energy wall.

In Figure 10a,b, for tests T1 and T2, respectively, the computed and measured exposure coefficients are plotted against the ratio between longitudinal velocity fluctuations obtained with approach B and the longitudinal double-averaged velocity. For the theoretical model, the following typical values were adopted: $\varphi = 30° \psi = 10°$, $\theta = 35°$, $C_D = 0.4$ and $C_D/C_L = 1.5$ (Dwivedi et al. [46]).

Figure 10 shows that there is little agreement between the derived and the measured exposure coefficients. There should be a very slight positive correlation as, for this sample, the expected value of u_p is lower when the exposure is very high. This means that, for the entrainment events detected in this study, the reference velocity defined as we did, is not useful to express threshold conditions and thus integrate bedload models based on assigning a probability to this threshold.

In our opinion, this line of research should not be abandoned but extra efforts will be needed to:

i.   inspect the performance of other definitions of reference velocity.

ii. place resources on the experimental characterization of drag and lift on sediment particles, taking into account local unsteadiness brought about by turbulence; the study of the inertia of the boundary layer should deserve some attention as this may have a strong impact on lift and drag coefficients.

iii. investigate how representative is entrainment due to fluid-particle momentum transfer, relatively to other forms of imparting momentum to bed particles, e.g., by particle-particle interactions.

iv. investigate in what other ways the flow field can be modified in the vicinity of the entrained bed particle without affecting the reference velocity measured above it.

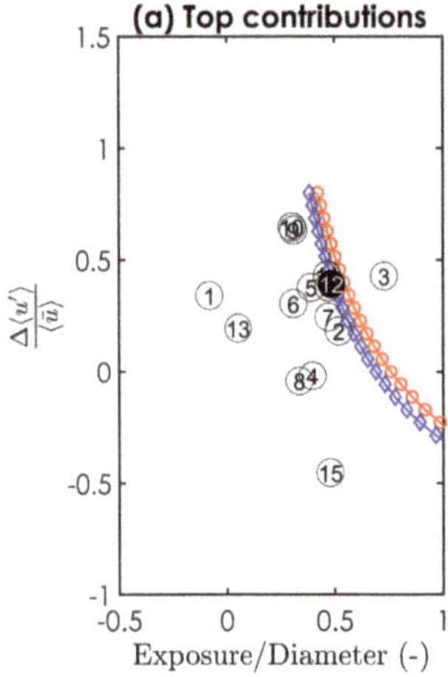
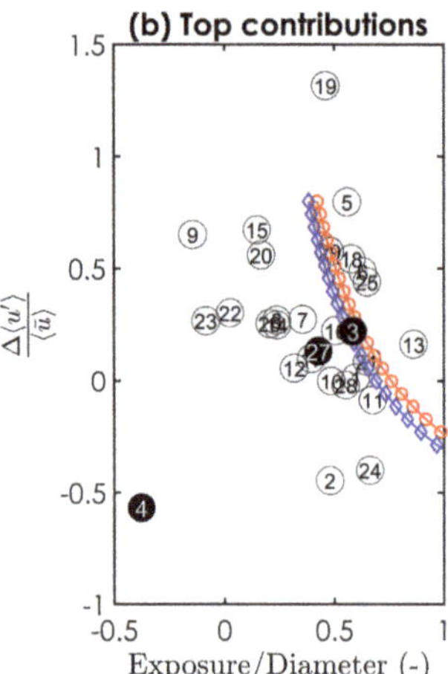

Figure 10. Ratio between longitudinal velocity fluctuations obtained with approach B and the longitudinal double-averaged velocities (averaged first in time and then in space-along the velocity reference level) as a function of the ratio between particle exposure and particle diameter for: (**a**) test T1; and (**b**) test T2. Each entrainment event corresponds to a numbered open circle. Curves corresponding to the theoretical exposure model proposed by Ferreira et al. [16] are preented as red circles model for sliding instability (Equation (6)) and as blue diamonds for rolling instability (Equation (7)).

In what concerns items iii. and iv., observations of the current database may help in devising future research paths. In particular, our observations allow for a closer scrutiny of each entrainment event, including the history of near-bed flow field and the typology of the entrainment, namely if occurred as a singular or a collective event [30]. A discussion of representative observations is presented in the next section.

4.3. Representative Types of Particle Entrainment

We consider that most entrainment events in our database could be grouped in four representative types:

A   singular events associated with non-locally generated hydrodynamic actions.
B   singular events associated with locally generated hydrodynamic actions.
C   collective entrainment events due to particle-particle momentum transfer collision.
D   collective entrainment associated with strong fluid flow events.

A description of specific examples is presented next.

4.3.1. Type A: Non-Local Hydrodynamic Actions

The first type considered responsible for particle entrainment has long been discussed in the literature, as already pointed out in Section 1. The extreme values of fluctuating turbulent forces are a key consideration in understanding particle entrainment. As seen before, particles can be entrained by a wide spectrum of hydrodynamic forces. Should entrainment occur at low velocities, the role played by the bed topography is surely significant. An example of particle dislodgement due to hydrodynamic forces is reported in Figure 11. This specific entrainment event belongs to Test T1 and is marked in Figure 7a,b with number 12.

Particle entrainment occurred with positive streamwise and vertical velocity fluctuations, respectively $u' = 0.1626$ m s^{-1} and $w' = 0.0484$ m s^{-1} and $u' = 0.1711$ m s^{-1} and $w' = 0.0307$ m s^{-1}, corresponding with outward interaction, although evidence of the presence of a sweep event in vicinity of the particle is observed in Figure 12, where the instantaneous Reynolds shear stresses, $u'w'$, computed for the same sequence of images of Figure 11 are reported.

An accurate inspection of all three plots shown in Figure 12, allowed us to reconstruct the history of this specific entrainment event. At time $t = t_0$ a sweep event ($u' > 0$ and $w' < 0$), is clearly identifiable in blue on the middle/top and left side of the image, is approaching. The remaining of that is observed in the next instant on the upstream side of the particle about to be dislodged.

Meanwhile a major recirculation area of negative Reynolds shear stresses is produced just downstream of the particle, resulting in dissipation of turbulent kinetic energy. On top of the particle positive streamwise and vertical velocity fluctuations are observed, as mentioned above. Although the turbulent structures seem to be not overly strong, particle entrainment is facilitated by the significant particle exposure (one particle radius, as detailed in Figure 10a) together with the configuration of the downstream neighbor. In the next instant the particle has already been dislodged and parcels of fluid characterized by intense Reynolds stresses (in blue) at the location previously occupied by the particle. At the instant before entrainment, as sweep approaches, the pore pressure increases. A few milliseconds later, the sweep passes over the particle and the pressure on top decreases; the combination of lift and direct drag causes particle dislodgement.

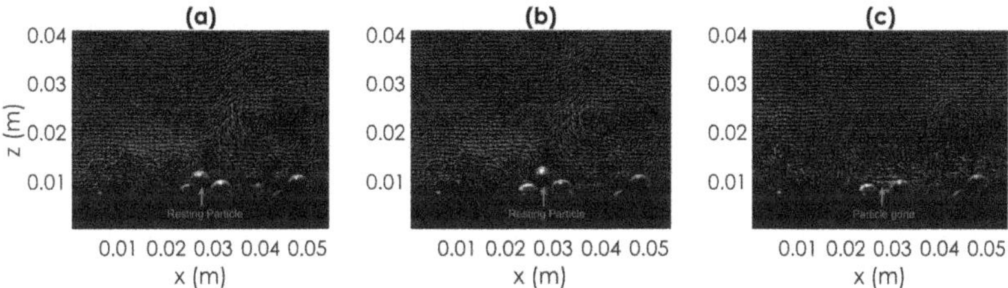

Figure 11. Particle entrained because of hydrodynamic forces (Type A; particle 12, Test T1). Sequence of three PIV images respectively showing (from (**a**–**c**)): at time $t = t_0$ (2 time instants, corresponding to 0.1333 s, before entrainment) the bright particle located in the centre of the image is at rest in the bed; at time $t = t_0 + \Delta t$ (1 time instant, corresponding to 0.067 s, before entrainment) the particle is still at rest in the bed; at time $t = t_0 + 2\Delta t$ the particle has already left the bed and is rolling. (For clarity only 1 out of 2 vectors are depicted.)

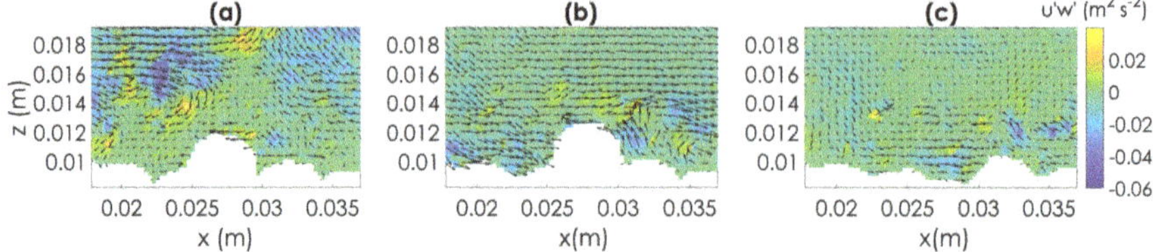

Figure 12. Instantaneous Reynolds shear stress maps with vectors superimposed representing velocity fluctuations obtained for Type A; (particle 12, Test T1) for the sequence of three PIV images reported in Figure 11. The sediment bed is masked in white. (For clarity only 1 out of 2 vectors are depicted.) (**a**) At time $t = t_0$ a sweep event ($u' > 0$ and $w' < 0$), is clearly identifiable in blue on the middle/top and left side of the image, is approaching. The remaining of that is observed in the next instant (**b**) on the upstream side of the particle about to be dislodged. In the next instant (**c**) the particle has already been dislodged and parcels of fluid characterized by intense Reynolds stresses (in blue) at the location previously occupied by the particle.

4.3.2. Type B: Locally Influenced Hydrodynamic Actions

Another mechanism observed in the PIV database consists in particle entrainment promoted by other sediment particles rolling or saltating nearby the particle at rest. The trajectory of the perturbing sediments is out of the plane of the laser sheet but close enough to disturb the flow field around the particle and cause its dislodgement. This mechanism may occur at low flow velocities, as the case of the particle removed from its rest position shown in Figure 13 and characterized by positive streamwise and negative vertical velocity fluctuations ($u' = 0.1429$ m s^{-1} and $w' = -0.0183$ m s^{-1} in approach A and $u' = 0.1198$ m s^{-1} and $w' = -0.0587$ m s^{-1} in approach B).

This event belongs to Test T2 and is marked in Figure 8a,b with number 3.

As for Type A, the instantaneous Reynolds shear stresses, $u'w'$, computed for the same sequence of images of Figure 13 are reported in Figure 14. In this case, PIV measurements in the plan Oxy (parallel to the bed) would have enabled a better understanding of the dynamics of the flow originated by the disturbing particle since it may have been the case that the particle may have been exposed to highly asymmetric actions that displaced it laterally to a position of greater exposure.

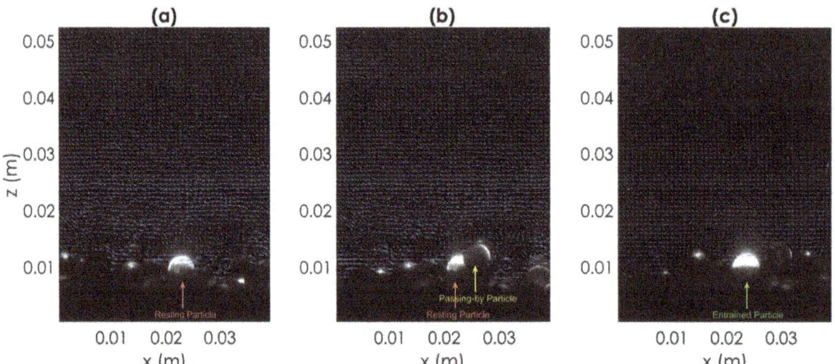

Figure 13. Particle entrained because of particle passing-by (Type B; Particle 3, Test T2). Sequence of three PIV images respectively showing (from (**a**–**c**)): at time $t = t_0$ (2 time instants, corresponding to 0.1333 s, before entrainment) the bright particle located in the centre of the image is at rest in the bed; at time $t = t_0 + \Delta t$ (1 time instant, corresponding to 0.067 s, before entrainment) the particle is still at rest in the bed and a particle perturbating the flow field passes nearby; at time $t = t_0 + 2\Delta t$ the particle starts its entrainment. (For clarity only 1 out of 2 vectors are depicted.)

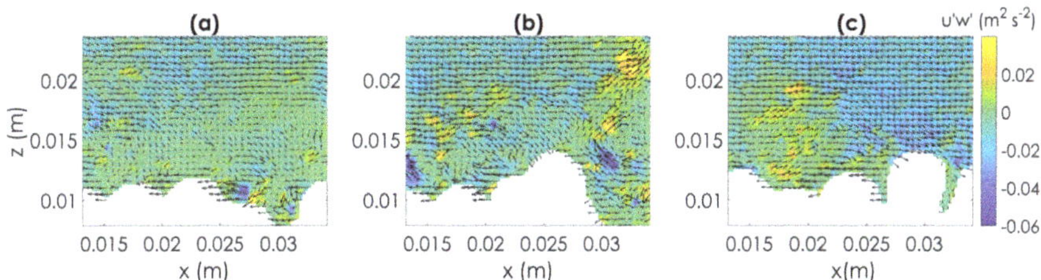

Figure 14. Instantaneous Reynolds shear stress maps with vectors superimposed representing velocity fluctuations obtained for Type *B* (Particle 3, Test T2) for the sequence of three PIV images reported in Figure 13. (For clarity only 1 out of 2 vectors are depicted.) (**a**) Instantaneous Reynolds stresses around particle at rest. (**b**) Flow disturbed by passing-by particle. (**c**) Instantaneous Reynolds stresses at the moment particle at rest is entrained.

More information can be provided analysing the flow velocity fluctuations obtained as the difference between instantaneous velocities and spatial average velocity (Figure 15) within the area represented in Figure 14 and corresponding instantaneous Reynolds shear stresses.

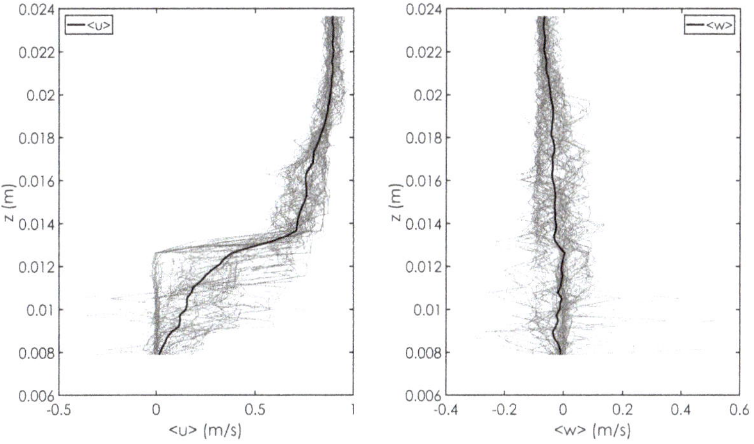

Figure 15. Instantaneous velocity profiles used to compute the spatial mean velocity (solid black line) in the region of interest (Figure 14).

4.3.3. Type C: Collective Entrainment Due to Particle Collision

The PIV data set includes several cases of particle dislodgement characterized by direct collisions between moving particles and particles at rest leading to the destabilization of the latter, especially in Test T2 where the bedload rate is much more significant.

This mechanism was already identified experimentally and in field surveys by Drake et al. [57], Böhm et al. [58] and investigated by Ancey et al. [30], who defined the entrainment of particles from the bed as the contribution of two processes: singular entrainment (at rate $\lambda_1 > 0$); collective entrainment, for instance associated with momentum of moving particles and transferred to resting particles (at rate μ).

Particle marked with number 27 in Figure 8a,b is considered hereafter as example of collective entrainment and depicted in Figure 16.

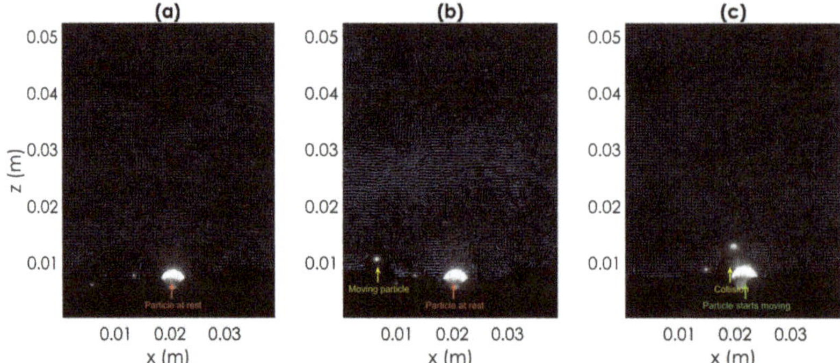

Figure 16. Particle entrained because of collective entrainment (Type C, Particle 27, Test T2). Sequence of three PIV images respectively showing (from (**a**–**c**)): at time $t = t_0$ (2 time instants, corresponding to 0.1333 s, before entrainment) the bright particle located in the center of the image is at rest in the bed; at time $t = t_0 + \Delta t$ (1 time instant, corresponding to 0.067 s, before entrainment) the particle is still at rest in the bed and another particle is approaching; at time $t = t_0 + 2\Delta t$ the travelling particle collides with the particle at rest; the bright particle starts its entrainment. (For clarity only 1 out of 2 vectors are depicted.)

From the instantaneous Reynolds shear stresses, $u'w'$, computed for the same sequence of images of Figure 16 and reported in Figure 17 it can be observed that no particular hydrodynamic event contributes to mobilise the particle and therefore the momentum imparted by the colliding particle seems the only cause responsible for particle entrainment. This means that the positive streamwise and vertical velocity fluctuations associated with this specific entrainment event shown in Figure 8c,d would not be sufficiently effective to pick up the particle without the extra momentum transmitted by sediment impact.

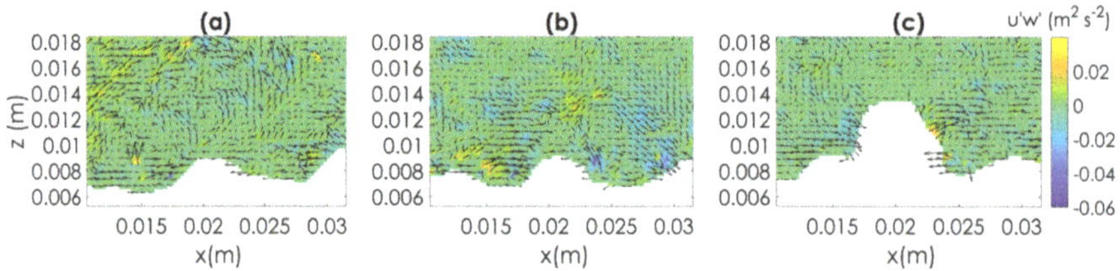

Figure 17. Instantaneous Reynolds shear stress maps with vectors superimposed representing velocity fluctuations obtained for Type C (Particle 27, Test T2) for the sequence of three PIV images reported in Figure 16. (For clarity only 1 out of 2 vectors are depicted.) (**a**) Instantaneous Reynolds stresses and particle at rest. (**b**) Instantaneous Reynolds stresses with particle at rest while moving particle is approaching. (**c**) Instantaneous Reynolds stresses during particles' collision.

4.3.4. Type D: Collective Entrainment Associated to Strong Fluid Flow Events

Although both types C and D are in this text denoted by *collective entrainment*, in the former the entrainment regards just a single particle and is generated by particle collisions, while in the latter dislodgement consists in simultaneous pickup of several sediments-characterized by a time scale varying between 0.067 s and 0.2 s (corresponding respectively with 1/15 s and 3/15 s) and occurs mainly because of two different causes:

1. Presence of high-speed gust mobilizing more than one particle at the same time.
2. Collisions between travelling particles and sediments at rest promoting the motion of the latter.

The entrainment of particle 4 of Test T2 belongs to the second situation. In Figure 18 the sequence of images representing this mechanism is reported, as well as the instantaneous Reynolds shear stresses, $u'w'$, in Figure 19.

In fact, the entrainment of particle 4 is related to ejections, which are negative longitudinal velocity fluctuations ($u' = -0.2321$ m s^{-1} with the first approach and $u' = -0.2846$ m s^{-1} with the second one) and positive vertical velocity fluctuations (respectively $w' = 0.1037$ m s^{-1} and $w' = 0.0527$ m s^{-1}).

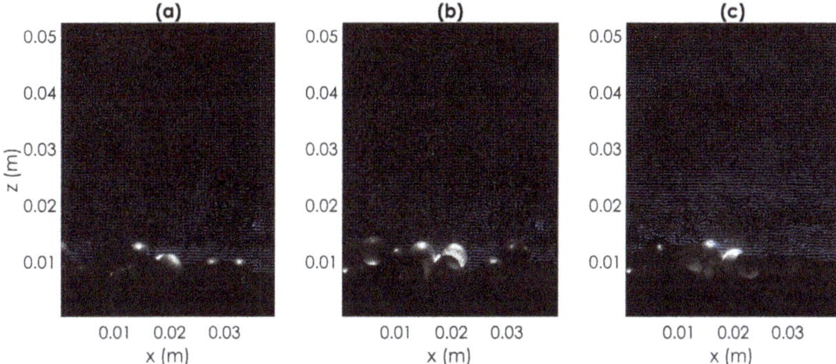

Figure 18. Particle entrained because of collective entrainment (Type D, particle 4, Test T2). Sequence of three PIV images respectively showing (from (**a**–**c**)): at time $t = t_0$ (2 time instants, corresponding to 0.1333 s, before entrainment) the bright particle located in the centre of the image is at rest in the bed (partially hidden by another particle); at time $t = t_0 + \Delta t$ (1 time instant, corresponding to 0.067 s, before entrainment) the particle is still at rest in the bed and travelling particles are about to collide with sediments in the bed; at time $t = t_0 + 2\Delta t$ several particles, included the bright one, are mobilized because of the impact. (For clarity only 1 out of 2 vectors are depicted.)

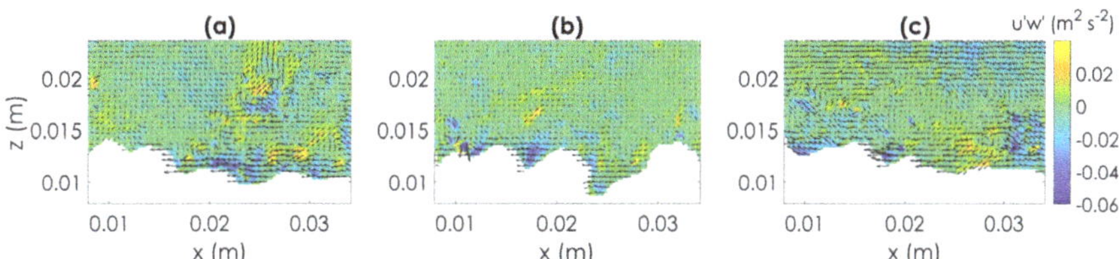

Figure 19. Instantaneous Reynolds shear stress maps with vectors superimposed representing velocity fluctuations obtained for Type D (particle 4, Test T2) for the sequence of three PIV images reported in Figure 18. (For clarity only 1 out of 2 vectors are depicted.) (**a**) Initial instantaneous Reynolds stresses. A ejection event is observed close to the particle at rest. (**b**) Induced instantaneous Reynolds stresses by passing particles. (**c**) Several particles are mobilized and corresponding instantaneous Reynolds stresses.

4.4. Disentrainment: Data Analysis and Results

At present little is known about causes and mechanics related with this disentrainment of bedload particles phenomenon. Disentrainment events are much more difficult to be identified either in field or in laboratory environments. In the present work, a visual inspection of the PIV footage allowed to select a sample of 11 sediment particles of test T1 going to deposit into the bed and analyse the flow field in their vicinity. The flow field related with disentrainment is assumed above the particle about to rest, namely the velocity vectors of the last frame before disentrainment in accordance with the second approach proposed in Section 4.3 for the case of sediment entrainment.

The quadrant plot obtained for disentrainment is depicted in Figure 20, where the instantaneous velocity fluctuations associated with particle disentrainment are represented with red dots.

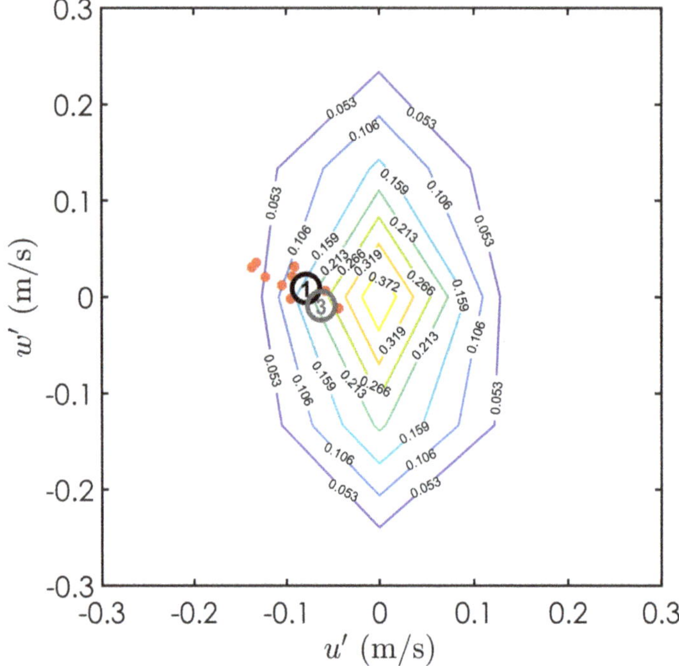

Figure 20. Instantaneous velocity fluctuations characterizing test T1—Approach B. Red dots represent the turbulent flow field associated with deposited particles. Disentrainment events discussed later are marked with numbered open circles: particle 1 (black) and particle 3 (gray).

Negative values of instantaneous velocity fluctuations in streamwise direction (u') are observed for all the observed events, while a larger range of turbulent velocities (positive and negative) can be noted for the w' component, corresponding to ejections and inward interactions. Ejections seem to be the prevalent flow state when disentrainment occurs. Only three of the registered events ocurred during inward interactions. This observation should be generalised with some care since the size of database is relatively small.

Relatively to entrainment, disentrainment events occur at al smaller range of along-wall velocities. Disentrainment seems to be much influenced by bed topography, as can be noted in the PIV footage: sediments in motion are more likely to become trapped within pockets if bed depressions are found along their path or if obstacles (bed particles particularly exposed) are responsible for their stop. This is in line with experimental findings of Cecchetto et al. [41].

Two disentrainment events among the sample of particles going at rest are hereafter analyzed. The sequence of images representing the disentrainment and associated instantaneous Reynolds stresses are reported in Figures 21–24, respectively for particle 1 (Figures 21 and 22) and particle 3 (Figures 23 and 24), whose velocity fluctuations are marked in Figure 20.

In the first case the particle in motion begins a deceleration phase until its complete rest due to the frictional contact with sediments constituting the bed. Particle disentrainment occurs when the inertial forces, that are momentum imparted by the fluid, pressure drag (depending on the relative velocity between particle and fluid) and viscous forces, are overcome by frictional contacts. Figure 22 shows the instantaneous Reynolds shear stresses next to particle 1 respectively at two and one instant before disentrainment (left and centre plots) and at the instant of rest (right plot). The fluctuation of the streamwise velocity is negative and the wall-normal is positive. This is thus a representative case of disentrainment during ejection, the most frequent state of near-bed fluid motion associated to disentrainment in our database.

In the second case the particle goes to rest occurs because of the presence of another sediment located along its path and enough exposed to constitute a barrier to the motion of particle 3, as observed in Figure 23. The obstacle provides force against motion and particle acceleration becomes abruptly to zero. Inward interaction is associated also with this event. Instantaneous Reynolds shear stresses are reported in Figure 24.

Disentrainment is associated with lower values of streamwise velocity fluctuations, although a key role in the process is played by the bed topography. Hydrodynamic forces alone do not determine the disentrainment of sediments in the bed. Local barriers or bed depressions or friction between rolling particles and sediment bed are the main causes related with this mechanism.

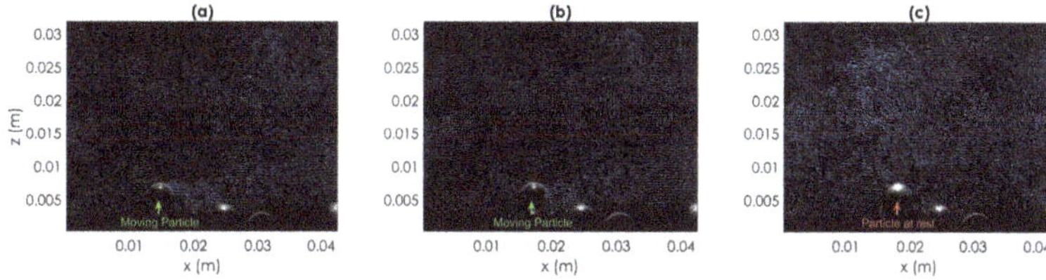

Figure 21. Particle deposited because of hydrodynamic forces (particle 1, Test T1). Sequence of three PIV images respectively showing (From (**a**–**c**)): at time $t = t_0$ (2 time instants, corresponding to 0.1333 s, before disentrainment) the particle is in motion within the field of view; at time $t = t_0 + \Delta t$ (1 time instant, corresponding to 0.067 s, before disentrainment) the particle is approaching its rest location; at time $t = t_0 + 2\Delta t$ the particle is at rest in the bed.

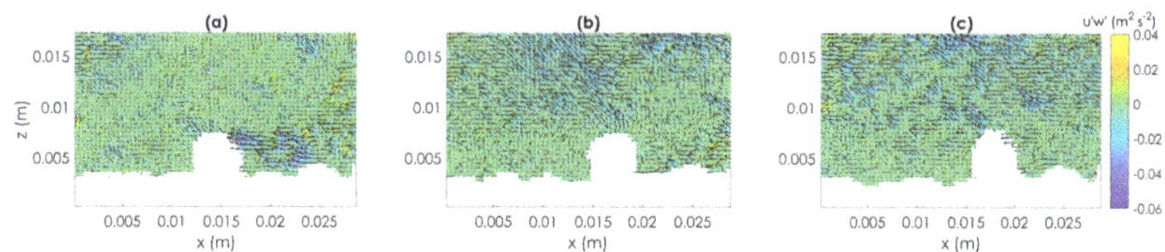

Figure 22. Instantaneous Reynolds shear stress maps with vectors superimposed representing velocity fluctuations obtained for disentrainment event number 1-Test T1 for the sequence of three PIV images reported in Figure 21. (**a**) the moving particle starts decelerating and a ejection event is observed downstream of the particle location. (**b**) The ejection event has passed and particle approaches its rest locatin. (**c**) The particle is at rest.

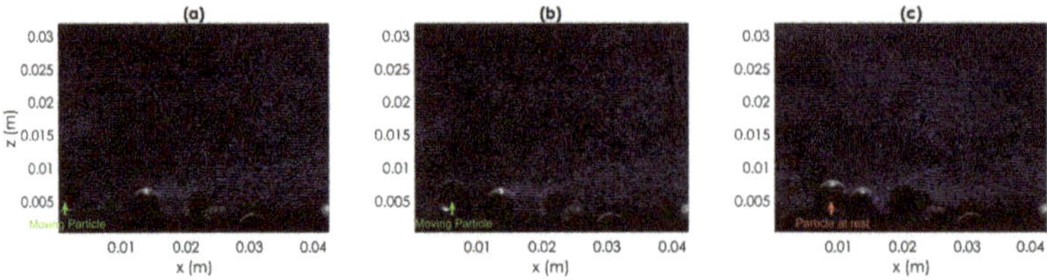

Figure 23. Particle deposited because of the presence of a cluster (particle 3, Test T1). Sequence of three PIV images respectively showing (From (**a**–**c**)): at time $t = t_0$ (2 time instants, corresponding to 0.1333 s, before disentrainment) the particle is in motion within the field of view; at time $t = t_0 + \Delta t$ (1 time instant, corresponding to 0.067 s, before disentrainment) the particle is approaching its rest location; at time $t = t_0 + 2\Delta t$ the particle is at rest in the bed.(For clarity only 1 out of 2 vectors are depicted.)

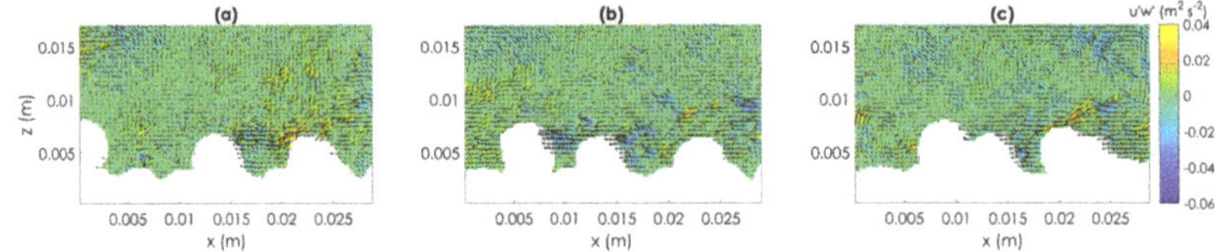

Figure 24. Instantaneous Reynolds shear stress maps with vectors superimposed representing velocity fluctuations obtained for disentrainment event number 3-Test T1 for the sequence of three PIV images reported in Figure 23. (For clarity only 1 out of 2 vectors are depicted.) (**a**) The moving particle enters the FoV. (**b**) As it aproaches a cluster of particles the instantaneous Reynolds shear stresse are significantly higher, as seen by the blue area close to the cluster. Ejection events are observed. (**c**) The moving particle stops due to the obstacle.

5. Conclusions

The experimental analysis reported in the present article aimed to investigate the kinematics of entrainment and disentrainment of uniform granular media subjected to a turbulent open-channel flow. Special consideration was given to the mechanisms promoting those events, as fluid-particle interactions, sediment collisions and the influence of the natural bed particle morphology.

The experimental program was designed to not compromise between fundamental features of sediment particles taking into account of the natural morphology of the sediment bed. The PIV technique was employed to characterize the longitudinal and vertical instantaneous turbulent fluctuations associated with sediment dislodgement and disentrainment and to determine a reference velocity above the particles crest.

Concerning particle entrainment, it was observed that sediment entrainment occurred at a wide range of turbulent flow velocities, with a prevalence of sweep and outward interactions. From our limited database it seems that entrainment may occur at velocities lower than average, but only in ejections events. From the same database, inward interaction events were almost not present.

A visual inspection of the PIV datasets enabled computing particle exposure and it was seen that an increase of particle exposure does not necessarily imply low flow velocities associated with particle dislodgement.

The factors involved in sediment entrainment are in fact multiple and several of these are not directly quantifiable, as the persistence of hydrodynamic actions or the influence of the downstream bed topography. Four types of particle entrainment were identified from the acquired PIV databases:

(i) the classic case of entrainment caused by hydrodynamic forces (Type A); (ii) the entrainment promoted by other sediments rolling or saltating nearby the particle at rest and therefore perturbing the flow field close to the particle and causing its dislodgement by imparting momentum transversally (Type B); (iii) sediment entrainment due to direct collisions between moving particles and those at rest in the bed (Type C) and (iv) entrainment due to simultaneous pickup of several sediment included the particle located under the plane of the laser sheet (Type D).

Cases (ii) and (iii) show the limitation of the reference velocity approach as a mean to determine the energy transfer from the flow to the particle, thus suggesting that further research is needed in the physics of flow-particle interaction in particular with respect to the drag and lif coefficients.

The flow field related with disentrainment events was analyzed as well: negative values of instantaneous velocity fluctuations in streamwise direction are observed for all the sample of particles, while both positive and negative vertical fluctuation components were found.

Bed topography also plays, in this case, a key role on the disentrainment events: sediments in motion were more likely to become trapped within pockets if bed depressions are found along their path or in presence of sediment barriers.

Author Contributions: Methodology: R.M.L.F.; bibliography research: F.A., R.M.L.F.; data acquisition: F.A. and A.M.R.; data processing software: R.A. Data Analysis: R.A., F.A., A.M.R., R.M.L.F.; writing and reviewing: R.A., F.A., A.M.R. and R.M.L.F. All authors have read and agreed to the published version of the manuscript.

Funding: This work was financially supported by: Project PTDC/ECM-HID/6387/2014–POCI-01-0145-FEDER-016825—funded by FEDER funds through COMPETE2020—Programa Operacional Competitividade e Internacionalização (POCI) and by national funds through FCT—Fundação para a Ciência e a Tecnologia, I.P. and partially supported by Portuguese and European funds, within the COMPETE 2020 and PORL-FEDER programs, through project RiverCure PTDC/CTA-OHR/29360/2017.

Conflicts of Interest: The authors declare no conflict of interest.

Abbreviations

The following abbreviations are used in this manuscript:

CFD	Computational Fluid Dynamics
FoV	Field of View
PIV	Particle Image Velocimetry
VLSM	Very Large-Scale Motion

References

1. Meyer-Peter, E.; Müller, R. Formulas for bedload transport. In Proceedings of the 2nd Meeting of the International Association for Hydraulic Research, Stockholm, Sweden, 7–9 June 1948.
2. Wong, M.; Parker, G. Reanalysis and Correction of Bed-Load Relation of Meyer-Peter and Mü ller Using Their Own Database. *J. Hydraul. Eng.* **2006**, *132*, 1159–1168 [CrossRef]
3. Smart, G.M. Sediment transport formula for steep channels. *J. Hydraul. Eng.* **1984**, *110*. [CrossRef]
4. Recking, A. An analysis of nonlinearity effects on bed load transport prediction. *J. Geophys. Res. Earth Surf.* **2013**, *118*, 1264–1281, [CrossRef]
5. Bagnold, R.A. *An Approach to the Sediment Transport Problem from General Physics*; USGS Numbered Series Professional Paper; U.S. Government Printing Office: Washington, DC, USA, 1966. [CrossRef]
6. Engelund, F.; Hansen, E. *A Monograph on Sediment Transport in Alluvial Stream*; Teknisk Forlag: Copenhagen, Denmark, 1967.
7. Rijn, L.C.V. Sediment transport part I bed load transport. *J. Hydraul. Eng.* **1984**, *110*. [CrossRef]
8. Wiberg, P.C.; Smith, J.D. Model for calculating bed load transport of sediment. *J. Hydraul. Eng.* **1989**, *115*. [CrossRef]
9. Einstein, H.A. Die eichung des im Rhein verwendeten geschiebefangers (The calibration of bed-load traps used in the Rhine). *Schweiz. Bauztg.* **1937**, *110*, 167–170.
10. Einstein, H.A. *The Bed-Load Function for Sediment Transportation in Open Channel Flows*; US Department of Agriculture: Washington, DC, USA, 1950; Volume 1026.

11. Lakatos, I. Falsification and the methodology of scientific research programmes. In *Can Theories Be Refuted? Essays on the Duhem-Quine Thesis*; Springer: Berlin/Heidelberg, Germany, 1976. [CrossRef]

12. Engelund, F.; Fredsoe, J. A sediment transport model for straight alluvial channels. *Hydrol. Res.* **1976**. [CrossRef]

13. Cheng, N.S.; Chiew, Y.M. Pickup Probability for Sediment Entrainment. *J. Hydraul. Eng.* **1998**, *124*. [CrossRef]

14. Yalin, M.S. *Mechanics of Sediment Transport*; Pergamon Press: Oxford, UK, 1972.

15. Dancey, C.L.; Balakrishnan, M.; Diplas, P.; Papanicolaou, A.N. The spatial inhomogeneity of turbulence above a fully rough, packed bed in open channel flow. *Exp. Fluids* **2000**, *29*, 402–410. [CrossRef]

16. Ferreira, R.M.; Hassan, M.A.; Ferrer-Boix, C. Principles of Bedload Transport of Non-cohesive Sediment in Open-Channels. In *Rivers-Physical, Fluvial and Environmental Processes*; Rowinsky, P., Radecki-Pawlick, A., Eds.; Springer: Berlin/Heidelberg, Germany, 2015; Chapter 13, pp. 323–372.

17. Paintal, A.S. A stochastic model of bed load transport. *J. Hydraul. Res.* **1971**, *9*, 527–554. [CrossRef]

18. Schmeeckle, M.W.; Nelson, J.M.; Shreve, R.L. Forces on stationary particles in near-bed turbulent flows. *J. Geophys. Res. Earth Surf.* **2007**, *112*. [CrossRef]

19. Valyrakis, M.; Diplas, P.; Dancey, C.L. Entrainment of coarse particles in turbulent flows: An energy approach. *J. Geophys. Res. Earth Surf.* **2013**, *118*, 42–53. [CrossRef]

20. Diplas, P.; Dancey, C.L.; Celik, A.O.; Valyrakis, M.; Greer, K.; Akar, T. The role of impulse on the initiation of particle movement under turbulent flow conditions. *Science* **2008**, *322*, 717–720. [CrossRef]

21. Chepil, W.S. Equilibrum of soil grains at the threshold of movement by wind. *Soil Sci. Soc. Am. J.* **1959**, *23*, 422–428. [CrossRef]

22. Leonardi, A.; Pokrajac, D.; Roman, F.; Zanello, F.; Armenio, V. Surface and subsurface contributions to the build-up of forces on bed particles. *J. Fluid Mech.* **2018**, *851*, 558–572 [CrossRef]

23. Rubey, W.W. *The Force Required to Move Particles on a Stream Bed*; Professional Papers; U.S. Geological Survey: Reston, VA, USA, 1938; pp. 121–141.

24. Bridge, J.S.; Dominic, D.F. Bed Load Grain Velocities and Sediment Transport Rates. *Water Resour. Res.* **1984**, *20*, 476–490. [CrossRef]

25. Papanicolaou, A.; Diplas, P.; Evaggelopoulos, N.; Fotopoulos, S. Stochastic incipient motion criterion for spheres under various bed packing conditions. *J. Hydraul. Eng.* **2002**, *128*, 369–380. [CrossRef]

26. Recking, A. A comparison between flume and field bed load transport data and consequences for surface-based bed load transport prediction. *Water Resour. Res.* **2010**, *46*, W03518. [CrossRef]

27. Ancey, C.; Bohorquez, P.; Heyman, J. Stochastic interpretation of the advection diffusion equation and its relevance to bed load transport. *J. Geophys. Res. Earth Surf.* **2015**, *120*, 2529–2551. [CrossRef]

28. Furbish, D.J.; Fathel, S.L.; Schmeeckle, M.W. Particle Motions and Bedload Theory. In *Gravel-Bed Rivers: Processes and Disasters*; Wiley: Hoboken, NJ, USA, 2017; pp. 97–120.

29. Houssais, M.; Ortiz, C.P.; Durian, D.J.; Jerolmack, D.J. Onset of sediment transport is a continuous transition driven by fluid shear and granular creep. *Nat. Commun.* **2015**, *6*, 1–8. [CrossRef]

30. Ancey, C.; Davison, A.; Böhm, T.; Jodeau, M.; Frey, P. Entrainment and motion of coarse particles in a shallow water stream down a steep slope. *J. Fluid Mech.* **2008**, *595*, 83–114. [CrossRef]

31. Cecchetto, M.; Tregnaghi, M.; Busolin, A.B.; Tait, S.; Marion, A. Statistical Description on theRole of Turbulence and Grain Interference on Particle Entrainment from Gravel Beds. *J. Hydraul. Eng.* **2017**, *143*, 06016021. [CrossRef]

32. Coleman, S.; Nikora, V.I. Fluvial dunes: Initiation, characterisation, flow structure. *Earth Surf. Process. Landf.* **2011**, *36*, 39–57. [CrossRef]

33. Furbish, D.J.; Haff, P.K.; Roseberry, J.C.; Schmeeckle, M.W. A probabilistic description of the bed load sediment flux: 1. Theory. *J. Geophys. Res. Earth Surf.* **2012**, *117*. [CrossRef]

34. Soares-Frazão, S.; Canelas, R.; Cao, Z.; Cea, L.; Chaudhry, H.M.; Moran, A.D.; el Kadi, K.; Ferreira, R.; Cadórniga, I.F.; Gonzalez-Ramirez, N.; et al. Dam-break flows over mobile beds: Experiments and benchmark tests for numerical models. *J. Hydraul. Res.* **2012**, *50*, 364–375. [CrossRef]

35. Andreotti, B.; Claudin, P.; Devauchelle, O.; Durán, O.; Fourriere, A. Bedforms in a turbulent stream: Ripples, chevrons and antidunes. *J. Fluid Mech.* **2011**, *690*, 94–128. [CrossRef]

36. Daubert, A.; Lebreton, J.C. Étude éxperimentale et sur modele mathematique de quelques aspects du calcul des processus d'erosion des lits alluvionaires en regime permanent et non permanent. In Proceedings of the 12th Congress of IAHR, Fort Collins, CO, USA, 11–14 September 1967; Volume 3, pp. 26–37.

37. Phillips, B.C.; Sutherland, A. Spatial lag effects in bed load sediment transport. *J. Hydraul. Res.* **1989**, *27*, 115–113. [CrossRef]

38. Charru, F.; Muilleron-Arnould, H.; Eiff, O. Erosion and deposition of particles on a bed sheared by a viscous flow. *J. Fluid Mech.* **2004**, *519*, 55–80. [CrossRef]

39. Canelas, R.; Murillo, J.; Ferreira, R.M.L. Two-dimensional depth-averaged modelling of dambreak flows over mobile beds. *J. Hydraul. Res.* **2013**, *51*, 392–407. [CrossRef]

40. Bohorquez, P.; Ancey, C. Stochastic-deterministic modeling of bed load transport in shallow water flow over erodible slope: Linear stability analysis and numerical simulation. *Adv. Water Resour.* **2015**, *83*, 36–54. [CrossRef]

41. Cecchetto, M.; Tait, S.; Tregnaghi, M.; Marion, A. The mechanics of bedload particles deposition over gravel beds. In Proceedings of the International Conference On Fluvial Hydraulics (River Flow 2016), St. Louis, MO, USA, 12–15 July 2016.

42. Ancey, C. Bedload transport: A walk between randomness and determinism. Part 2. Challenges and prospects. *J. Hydraul. Res.* **2020**, *58*, 18–33. [CrossRef]

43. Kramer, H. Sand mixtures and sand movement in fluvial model. *Am. Soc. Civ. Eng.* **1935**, *100*, 873–878.

44. Mendes, L.; Antico, F.; Sanches, P.; Alegria, F.; Aleixo, R.; Ferreira, R.M. A particle counting system for calculation of bedload fluxes. *Meas. Sci. Technol.* **2016**, *27*, 125305. [CrossRef]

45. Ferreira, R.M.L. The von Kármán constant for flows over rough mobile beds. Lessons learned from dimensional analysis and similarity. *Adv. Water Resour.* **2015**, *81*, 19–32. [CrossRef]

46. Dwivedi, A.; Melville, B.W.; Shamseldin, A.Y.; Guha, T.K. Analysis of hydrodynamic lift on a bed sediment particle. *J. Geophys. Res. Earth Surf.* **2011**, *116*. [CrossRef]

47. Antico, F. Laboratory Investigations on the Motion of Sediment Particles in Cohesionless Mobile Beds under Turbulent Flows. Ph.D. Thesis, Instituto Superior Técnico, Universidade de Lisboa, Lisboa, Portugal, 2019.

48. Nakagawa, H.; Nezu, I. Prediction of the contributions to the Reynolds stress from bursting events in open-channel flows. *J. Fluid Mech.* **1977**, *80*, 99–128. [CrossRef]

49. Nikora, V.; Ballio, F.; Coleman, S.; Pokrajac, D. Spatially averaged flows over mobile rough beds: Definitions, averaging theorems, and conservation equations. *J. Hydraul. Eng.* **2013**, *139*, 803–811. [CrossRef]

50. Kim, K.C.; Adrian, R.J. Very large-scale motion in the outer layer. *Phys. Fluids* **1999**, *11*, 417–422. [CrossRef]

51. Schoklitsch, A. *Handbuch des Wasserbaues*; Springer: Berlin/Heidelberg, Germany, 1962.

52. Gyr, A.; Schmid, A. The different ripple formation mechanisms. *J. Hydraul. Res.* **1989**, *27*, 61–74. [CrossRef]

53. Séchet, P.; Guennec, B.L. Bursting phenomenon and incipient motion of solid particles in bed-load transport. *J. Hydraul. Res.* **1999**. [CrossRef]

54. Nelson, J.M.; Shreve, R.L.; McLean, S.R.; Drake, T.G. Role of near-bed turbulence structure in bed load transport and bed form mechanics. *Water Resour. Res.* **1995**, *31*, 2071–2086. [CrossRef]

55. Ancey, C.; Böhm, T.; Jodeau, M.; Frey, P. Statistical description of sediment transport experiments. *Phys. Rev. E* **2006**, *74*, 011302. [CrossRef] [PubMed]

56. Valyrakis, M.; Diplas, P.; Dancey, C.L.; Greer, K., Celik, A.O. Role of instantaneous force magnitude and duration on particle entrainment. *J. Geophys. Res. Earth Surf.* **2010**, *115*. [CrossRef]

57. Drake, T.G.; Shreve, R.L.; Dietrich, W.E.; Whiting, P.J.; Leopold, L.B. Bedload transport of fine gravel observed by motion-picture photography. *J. Fluid Mech.* **1988**, *192*, 193–217. [CrossRef]

58. Böhm, T.; Ancey, C.; Frey, P.; Reboud, J.L.; Ducottet, C. Fluctuations of the solid discharge of gravity-driven particle flows in a turbulent stream. *Phys. Rev. E* **2004**, *69*, 061307. [CrossRef] [PubMed]

Article

Hydrodynamic Structure with Scour Hole Downstream of Bed Sills

Mouldi Ben Meftah *, Francesca De Serio, Diana De Padova and Michele Mossa

Department of Civil, Environmental, Land, Building Engineering and Chemistry, Polytechnic University of Bari, Via E. Orabona 4, 70125 Bari, Italy; francesca.deserio@poliba.it (F.D.S.); diana.depadova@poliba.it (D.D.P.); michele.mossa@poliba.it (M.M.)

* Correspondence: mouldi.benmeftah@poliba.it; Tel.: +39-080-5963-508

Received: 2 December 2019; Accepted: 6 January 2020; Published: 9 January 2020

Abstract: Experimental turbulence measurements of scour hole downstream of bed sills in alluvial channels with non-cohesive sediments are investigated. Using an Acoustic Doppler Velocimeter (ADV), the flow velocity-field within the equilibrium scour hole was comprehensively measured. In this study, we especially focus on the flow hydrodynamic structure in the scour hole at equilibrium. In addition to the flow velocity distribution in the equilibrium scour hole, the turbulence intensities, the Reynolds shear stresses, the turbulent kinetic energy, and the turbulent length scales are analyzed. Since the prediction of equilibrium scour features is always very uncertain, in this study and based on laboratory turbulence measurements, we apply the phenomenological theory of turbulence to predict the maximum equilibrium scour depth. With this approach, we obtain a new scaling of the maximum scour depth at equilibrium, which is validated using experimental data, satisfying the validity of a spectral exponent equal to −5/3. The proposed scaling shows a quite reasonable accuracy in predicting the equilibrium scour depth in different hydraulic structures.

Keywords: scour; velocity field; turbulence; equilibrium scour depth; new scaling of scour depth

---

1. Introduction

Prediction of maximum scour depth downstream of hydraulic structure i.e., bridge piers and abutments, sills, sluice gates, spillways, weirs, offshore platforms, wind turbines, etc., is of primary concern for a wide range of engineering and environmental applications. This topic has drawn attention and interest from many researchers for decades [1–14]. Despite these numerous studies, prediction of equilibrium-scour hole characteristics always remains challenging because of the complexity of the phenomenon and its dynamic sensitivity to structure and sediment properties. Most of these studies [3–9,12–14] proposed different empirical formulae based on experimental/field measurements, to predict the maximum eroded depth, its maximum length, and other properties. Ben Meftah and Mossa [3], Tregnaghi et al. [14], and Lu et al. [15] observed that, based on laboratory measurements of steady/unsteady flows, the scour downstream of a grade control structure evolves into three distinct phases, including an initial phase, a developing phase, and an equilibrium phase. Tregnaghi et al. [14] argued that the scour process usually reaches its equilibrium condition rapidly in live-bed conditions and rather slowly in clear-water conditions. Lu et al. [15] indicated that the scour hole in non-cohesive sediments is influenced by both the channel characteristics and the sediment properties, especially the channel bed slope, the densimetric Froude number, the tailwater depth, and the sediment median size.

The enormous amount of studies conducted on this issue asserts that the scour hole profiles are similar in shape, giving rise to a typical profile with appropriate scaling of the horizontal and vertical coordinates. However, in spite of the great effort made by researchers, many different formulae were derived to predict the scour profile at equilibrium. This large number of different empirical formulae, sometimes composed of complicated parameters, makes them less operational in practice than expected. Moreover, most of these formulae are affected by large uncertainties and suffer some limitations. According to Manes and Brocchini [16], the approaches based on dimensional analysis suffer from two main shortcomings: One due to the experimental laboratory scale issues, hiding the real shape of functional relations between non-dimensional groups at field scales, and the other one is related to the fact that the empirical approach does not provide a theoretical framework to interpret the experimental data and to understand the physics underlying such functional relations.

Recent studies [16–19] proposed very important and innovative approaches to predict localized turbulent flow scours, applying the phenomenological theory of turbulence (PTT). This approach hypothesizes that the scour process is controlled by the momentum transport generated by eddies belonging to the dissipation and production spectrum ranges. By scaling the eddy-characteristic-lengths of these spectrum-ranges with the equilibrium-scour dimensions and the characteristic-sediment-length, researchers tried to derive general predictive formulae, by merging the PTT-theoretical aspects with empirical observations, for the equilibrium maximum-scour depth at some hydraulic structure.

The main aim of this study is to contribute to this novel kind of approach, being the prediction of the scour features based on experiments and theory still challenging due to the complexity of the phenomenon. Therefore, in the present study, we first experimentally focus on the flow turbulence measurements in a scour hole developed downstream of a grade control structure in sand-bed channels, providing an integrated hydrodynamic picture of the scouring process. Successively, we propose a new scaling of the maximum scour depth at equilibrium and validate it using the experimental data of this study and some data collected from previous studies. Specifically, the proposed scaling approach is easy to use, depending in particular on a densimetric Froude number and on a relative roughness. Nevertheless, its application shows a quite reasonable accuracy in predicting the equilibrium scour depth in different hydraulic structures. Therefore, our findings would contribute to improve the understanding of the scouring mechanisms by applying the phenomenological theory of turbulence.

2. Experimental Set-Up

The experiments on the scour processes were carried out in a rectangular flume of closed-circuit flow at the Hydraulic Laboratory of the Mediterranean Agronomic Institute of Bari (Italy). The flume has glass sidewalls and a Plexiglass floor, allowing a good side view of the flow. It is 7.72 m long, 0.30 m wide, and 0.40 m deep. A pump of maximum discharge of 24 l/s was used to deliver water from the laboratory sump to an upstream tank equipped with a baffle and lateral weir, maintaining a constant head upstream of a movable slide-gate constructed at the inlet of the flume. The slide-gate regulates channel flow-discharge. To create a smooth flow transition from the upstream reservoir to the flume, a wooden ramp was installed at the inlet of the flume; it is 1.55 m long, 0.15 m thick and of same channel width (Figure 1). At the outlet of the flume, water is intercepted by a stilling tank, equipped with three vertical grids to stabilize water, and a triangular weir (V-notch sharp crested weir) to measure discharge with relative uncertainty of ±8%. At the downstream end of the flume, a movable gate made of Plexiglass and hinged at the channel bottom is used to regulate the flow depth.

In order to simulate grade control structures protecting riverbeds against erosion, in this study we have used a series of sills consisting of PVC plates 0.30 m wide and 0.01 m thick. The sills were installed on an experimental area extended 6 m along the channel, downstream of the wooden ramp. The sill height decreases progressively going downstream from the wooden ramp, respecting a determined initial slope S_0 of 0.0086. Different configurations were investigated, the difference between them being the distance, L, between sills. More details on the sills distribution are reported in Ben Meftah and Mossa [3].

The flume bottom downstream of the wooden ramp and between the sills is covered with an erodible bed material layer, consisting of almost uniform sand particles with a mean average size, d_{50}, of 1.8 mm and density of 2650 kg/m^3 (see Ben Meftah and Mossa [3] for more information). Along the experimental area, the sand layer was leveled respecting the maximum sill heights and that of the upstream wooden ramp, forming the original bed of the channel with a slope S_0 (Figure 1).

Front view

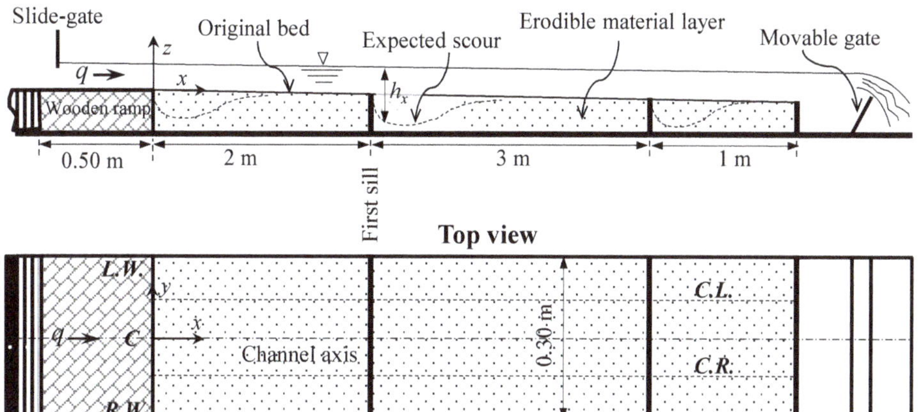

Figure 1. General sketch of the laboratory flume with the initial condition and expected scour hole (dashed profile) downstream of bed sills. h_x = flow depth in the expected equilibrium scour at a downstream position x, L.W. = left wall side of channel, R.W. = right wall side, C = centerline (channel axis), C.L. = centerline of left half of channel, and C.R. = centerline of right half, (x, y, z) = longitudinal, transversal, and vertical directions, respectively.

The data collected during each test included discharge, water surface elevation, flow depth, temporally eroded bed profile, equilibrium bed profile, and scour dimensions (depth, length, position of maximum depth). The profiles of the equilibrium eroded bed along the channel centerline, near the channel sidewalls, and at an intermediate distance between the channel centerline and both sidewalls (Figure 1) were measured, as the vertical distance between the initial bed elevation and the bed at equilibrium, by means of a point gauge of ±0.1 mm accuracy. The water level profile along the channel centerline was measured using an electrical hydrometer with an accuracy of ±0.1 mm.

In addition to the measurements of the scour geometric characteristics, the flow velocity-fields were also carried out in the scour hole at equilibrium condition. The velocity data were collected using a 3D Acoustic Doppler Velocimeter (ADV) system, developed by Nortek, with a sampling rate of 25 Hz at a time window of 70 s. The sampling volume of the ADV was located 5 cm below the transmitter probe. The ADV was used with a velocity range equal to ±0.30 m/s, a measured velocity accuracy of ±1%, and a sampling volume of less than 0.25 cm^3. For high-resolution measurements, the manufacturer recommends a 15 db signal-to-noise ratio (SNR) and a correlation coefficient larger than 70%. The acquired data were filtered based on the Tukey's method and bad samples (SNR < 15 db and correlation coefficient < 70%) were also removed. Additional details concerning the ADV-system operations can be found in [20–27]. Flow velocity measurements through the scour hole were carried out for different configurations in both the longitudinal plane of symmetry and in some transversal planes.

The initial experimental conditions and the geometric characteristics of scours, related to this study, are illustrated in Table 1, where h_c is the flow depth over the crest of the sill downstream of which the scour hole is measured, U_c is the flow velocity over the sill (mean velocity in correspondence of h_c), z_s is the maximum equilibrium scour depth from the original bed profile, h_s is the flow depth at the position of maximum equilibrium scour depth, $\lambda_c = d_{50}/h_c$ is a relative roughness, $F_{dc} = U_c/(\Delta g d_{50})^{0.5}$ is the densimetric Froude number for the approach flow over the sill, $\Delta = [(\rho_s - \rho_w)/\rho_w]$ is the submerged relative density of sediment particles, ρ_w is the water density, ρ_s is the sediment density, g is the gravitational acceleration, $Re_c = U_c h_c/\nu$ is the Reynolds number for the approach flow over the sill, $Re_g = U_c d_{50}/\nu$ is the grain Reynolds numbers, and ν is the water kinematic viscosity. For the sake of brevity, in this study we focus in detail on data of run T21 (Table 1) for the analysis of the turbulent parameters, while we adopt all runs (T04–T22) in the scaling procedure.

Table 1. Initial experimental conditions and parameters of the investigated runs.

Runs	L (m)	h_c (m)	U_c (m/s)	z_s (m)	h_s (m)	F_{dc} (-)	λ (-)	Re_c (-)	Re_g (-)
T04	1	0.035	0.593	0.039	0.10	3.474	0.051	18,159	934
T05	1	0.048	0.689	0.088	0.15	4.035	0.038	28,517	1069
T06	1	0.029	0.522	0.028	0.08	3.060	0.062	12,497	776
T07	1	0.054	0.726	0.105	0.18	4.255	0.033	32,354	1078
T08	1	0.042	0.647	0.064	0.13	3.790	0.043	23,774	1019
T09	2	0.036	0.585	0.053	0.09	3.425	0.050	18,942	947
T10	2	0.042	0.634	0.057	0.12	3.717	0.043	25,981	1113
T11	2	0.048	0.688	0.070	0.13	4.033	0.038	32,217	1208
T12	2	0.054	0.736	0.081	0.15	4.312	0.033	39,743	1325
T13	2	0.060	0.762	0.090	0.17	4.466	0.030	46,880	1406
T14	4	0.034	0.576	0.076	0.09	3.374	0.053	20,563	1089
T15	4	0.042	0.647	0.090	0.12	3.790	0.043	29,208	1252
T16	4	0.048	0.691	0.112	0.14	4.048	0.038	37,309	1399
T17	4	0.030	0.533	0.065	0.08	3.125	0.060	18,000	1080
T18	3	0.029	0.531	0.050	0.07	3.113	0.062	17,721	1100
T19	3	0.036	0.586	0.061	0.10	3.433	0.050	24,784	1239
T20	3	0.042	0.653	0.071	0.12	3.827	0.043	32,236	1382
T21	3	0.048	0.686	0.084	0.14	4.019	0.038	38,688	1451
T22	3	0.052	0.762	0.094	0.17	4.463	0.035	46,543	1611

3. Results and Discussion

3.1. Velocity Fields

Since turbulence is the most important mechanism of sediment entrainment, causing a significant increase in the shear stress around the base of a hydraulic structure, a large set of measurements of the flow velocity field in the scour holes was carried out. Figure 2, as an example, shows a vector map of the flow velocity, V_{xz}, in the scour hole downstream of the first bed sill, located 2 m downstream of the wooden ramp (Figure 1). In Figure 2 the (x, z)-coordinates take origins at the first sill position and the channel bottom, respectively. The V_{xz}-velocity is the resultant of the streamwise U and vertical W time-averaged velocity components. The three profiles of the initial bed (solid line), the free-surface flow (triangle down), and the bed at equilibrium condition (bullet) are also reported in Figure 2. The random point cloud in Figure 2 represents the remaining amount of sediment between sills at scour-equilibrium. All the data illustrated in Figure 2 were obtained in the plane of flow-symmetry $(y = 0)$.

Figure 2 clearly shows the flow velocity behavior through the scour hole. Three flow velocity distribution regions can be recognized in Figure 2: (i) A first region, 1, where a sort of a free entering jet flows, originated by the flow condition over the sill-crest; (ii) a second region, 2, characterized by vortex formations (eddies) due to the jet diffusion, located near the bottom of the scour hole and extended along the upstream scour-side; and (iii) a third region, 3, seeming less turbulent and taking place downstream, outside the vortex region. Between the regions 1 and 2, a sort of a hydraulic jump may occur, depending on the hydraulic conditions. The absence of velocity measurements in the upper flow region is due to the limitation of the ADV-downlooking probe, being the uppermost 7 cm of the flow could not be sampled. However, in the jet-like region 1, the acquired ADV-signal was very noisy, which could be due to the strong jet-flow agitation interacting with the hydraulic jump.

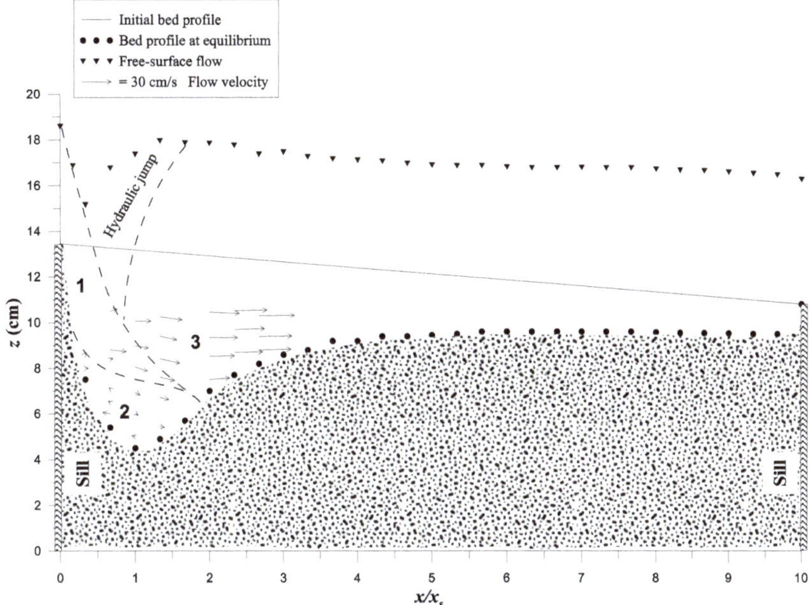

Figure 2. Vector map of the flow velocity, V_{xz}, in the scour hole at the plane of flow-symmetry ($y = 0$). The dashed line qualitatively indicates the separation between regions 1, 2, and 3.

The jet-like flow (region 1) plays a crucial role in the different phases of the scour development. This is due to its high velocity, which leads to an increase of the jet potential erosive action on the bed channel. As the jet size increases over time, the jet begins to gradually lose its erosive potential. The state of equilibrium occurs when the path of the impinging jet becomes sufficiently long and its diffused velocity is reduced to values lower than the minimum value required for sediment movement [28].

In the region 2 (Figure 2), a significant reduction of the flow velocity occurs. Furthermore, the flow distribution shows two portions: A portion of negative velocity starting at the position of maximum scour depth and extending towards the upward bed sill, forming a sort of clockwise local vortex, and another portion of positive velocity that shifts the flow downstream, from the position of the maximum scour depth. Similar behaviors have been observed by Ghodsian et al. [29]. The authors named as "weak flow" the portion of negative velocity and "strong flow" the portion of positive velocity. They also observed that the lowest region of the scour hole is mainly covered by the sediment of size d_{90}, grain size for which 90% of sampled particles are finer.

In the region 3 (Figure 2), the flow velocity considerably increases, compared to region 2. This increase is gradual in the downstream direction. Moreover, the velocity vectors tend to be more horizontal and of almost comparable values. The flow redistribution in region 3 indicates a sort of smooth transition flow from scour hole to the downstream tailwater flow. This smooth transitional flow result in less flow turbulence, which is the subject of the next section.

3.2. Turbulence Intensity Associated with Scour Hole

To get further information on the scouring process, in Figure 3 we plot the streamwise turbulence intensity, U', as a function of the normalized vertical coordinate Z/z_s. Herein, U' was defined as the ratio of the standard deviation of the streamwise flow-velocity component fluctuations to the average velocity, U_c, measured over the sill-crest, Z is the vertical position from the original bed profile (solid line in Figure 2) at a given downstream position x. The vertical U'-profiles correspond to different downstream positions $x/x_s = 0.33, 0.67, 1.00, 1.33, 2.00, 2.33,$ and 2.67, where x_s is the x-position from the grade-control structure (sill) at which the scour attains its maximum depth. Note that, due to the flow symmetry, the spanwise velocity, V, is theoretically expected to be null and therefore it has not any physical significance in this plane.

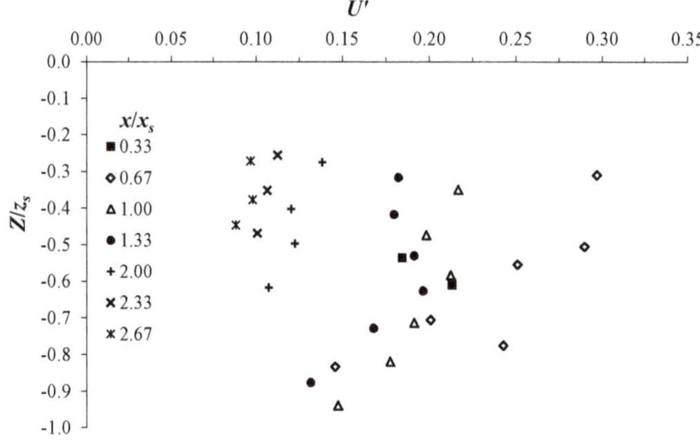

Figure 3. Vertical profiles of streamwise turbulence intensity, U', at different downstream positions x/x_s along the plane of flow symmetry ($y = 0$).

Figure 3 shows that the maximum turbulence intensities take place at $x/x_s \leq 1.33$, at the regions 1, 2 and the upstream side of region 3. At these regions, which practically occupy the whole part of the main scour hole, the values of turbulence intensities range between a minimum of 0.13 and a maximum of 0.3. At $x/x_s = 0.67$, U' experiences the maximum measured values at $Z/z_s \geq -0.7$. This position is located within the region 1, where the jet flow penetrates into the scour pool, generating high levels of turbulence. Figure 3 mainly indicates a tendency of U' to reduce as going down towards the scour bed. In region 2, at $Z/z_s < -0.7$, U' reduces by almost 50% as compared to region 1. The significant reduction in turbulence intensity in region 2 is related to the sharp decrease in flow velocity in this region, as shown in Figure 2. At the exit from the scour hole, at $x/x_s > 1.33$ in region 3, U' shows the smallest values, as expected based on the flow velocity distribution (Section 3.1). At this region, U' decreases almost twice compared to region 1 and by 60% compared to region 3.

In Figure 4 we plot the vertical turbulence intensity, W', profiles at the same downstream positions $x/x_s = 0.33, 0.67, 1.00, 1.33, 2.00, 2.33$, and 2.67. Herein, W' is defined as the ratio of the standard deviation of the vertical flow-velocity component fluctuations to U_c. Figure 4 clearly shows a substantial reduction in W' as compared to U'. For all the profiles, W' ranges between a minimum of 0.03 and a maximum of 0.17 against 0.13 and 0.30, respectively, observed for U'. Furthermore, at the different downstream positions x/x_s, W' decreases with decreasing Z/z_s (i.e., going down to the equilibrium-scour bed). This decrease is continuous and with a significant reduction rate as compared to U'. This may be explained by the considerable reduction of the vertical velocity flow going towards the equilibrium bed profile, which effectively reduces the vertical flow-force that could lift the sediments from the bottom. Similar to U', W' experiences maximum values in region 1 and at the upstream side of region 3. In region 2 however, W' exhibits the smallest values, comparable to those that occurred in region 3.

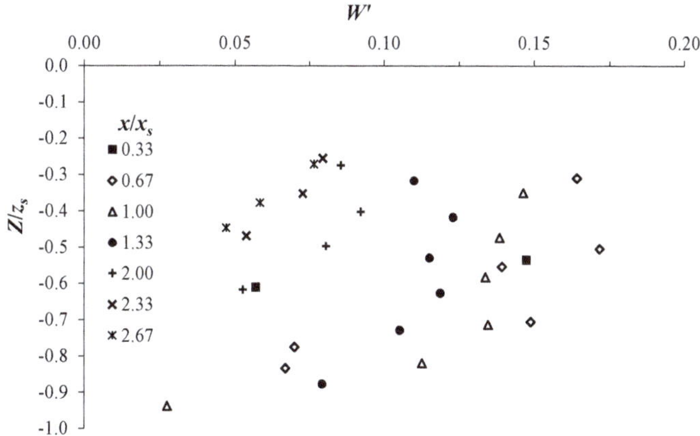

Figure 4. Profiles of vertical turbulence intensity, W', at different downstream positions x/x_s along the plane of flow symmetry ($y = 0$).

Figure 5 illustrates the vertical profiles of the normalized Reynolds shear stresses in the scour hole. The profiles were taken at the downstream positions $x/x_s = 0.33, 0.67, 1.00, 1.33, 2.00, 2.33$, and 2.67. In Figure 5, $U'W' = -<u'w'>/U_c^2$, where $<u'w'>$ is the time-averaged stress over the length of the time series and (u', w') are the velocity fluctuations of the streamwise and vertical component, respectively. The values of $U'W'$ clearly have a heterogeneous distribution at the different downstream positions x/x_s. This heterogeneity vertically decreases downward (towards the scour bed). Furthermore, Figure 5 shows that the heterogeneity of the Reynolds stresses is spatially variable in the scour hole. At $x/x_s = 0.67$, $U'W'$ varies from a maximum value $O(4 \times 10^{-2})$ to a value $O(10^{-3})$. Both the magnitude and the variation range of $U'W'$ gradually decrease with increasing x/x_s. For $x/x_s > 1.33$, there is a sharp decrease of the Reynolds stress magnitude and the values of $U'W'$ tend to an homogeneous distribution, from a maximum predictable averaged value $O(4 \times 10^{-3})$ to a value $O(10^{-3})$. Figure 5 also shows that $U'W'$ exhibits the largest values in region 1 (see Figure 2), it decreases slightly in the upstream side of region 3, and it is significantly reduced in region 2. At the edge of the downstream scour-side, $x/x_s > 1.33$, $U'W'$ shows the lowest values.

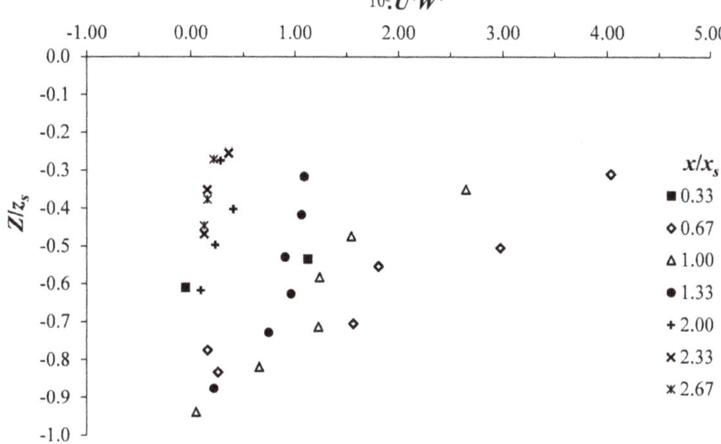

Figure 5. Vertical profiles of normalized Reynolds shear stress, $U'W'$, in scour hole and at different downstream positions x/x_s along the plane of flow symmetry ($y = 0$).

Figure 5 also indicates that $U'W'$ always has positive values in the scour hole, along the plane of flow symmetry ($y = 0$). This reflects a clear idea upon the frictional drag distribution and the vertical effective momentum transfer associated with the scour hole structure in the plane of flow symmetry. In the scour hole, the Reynolds shear stress is developed due to the formation of eddies of many different length scales. In the immediate vicinity of the sediment scour bed, localized turbulent eddies play an important role in removing a sediment particle from its stabilized position. According to Ali and Dey [30], at the flow-bed interface, the eddies of greater size than the sediment diameter, d_{50} as an example, weakly contribute to the vertical velocity component. By contrast, the eddies of smaller size than the sediment diameter perfectly fit in the space between the particles, providing an effective contribution to the vertical velocity component and therefore a substantial transfer of the vertical momentum may occur. The increase of the vertical lift force generated by small eddies (with smaller size than the sediment diameter) and the important horizontal component of momentum transmitted by large eddies (with greater size than the sediment diameter) at the sediment bed, play a crucial role in putting the sediment particles in suspension. Applying the phenomenological theory of turbulence and a dimensional analysis, Ali and Dey [30] found a scale of the Reynold shear stress at the bed for the incipient motion of sediment particles.

$$-\rho_w \langle u'w' \rangle \big|_b \sim \rho_w U_b{}^2 \lambda^{-\frac{(1+\sigma)}{2}} \tag{1}$$

where the subscript b indicates the flow-bed interface position, U_b is the threshold velocity, defined as the near-bed velocity that is marginally sufficient to initiate the particle motion at the bed surface, $\lambda (= d_{50}/h_x)$ is a relative roughness, h_x is defined in Figure 1, σ is the spectral exponent of the turbulent energy spectrum. By equating the Reynolds shear stress, obtained near the bed surface, and the bed shear stress τ_b, we can obtain the threshold velocity U_b. τ_b can be related to the gravitational shear stress τ_g as $\tau_b \sim \tau_g = (\rho_s - \rho_w) g d_{50} \theta_b$, where θ_b is the threshold Shields parameter. θ_b is a function of a particle parameter $D = [(g \Delta d_{50})/\nu^2]^{1/3}$. Combining the different parameters together, one obtains a scaling expression of the threshold densimetric Froude number, F_{db}:

$$\rho_w U_b{}^2 \lambda^{-\frac{(1+\sigma)}{2}} \sim (\rho_s - \rho_w) g d_{50} f(D)$$
$$F_{db} = \frac{U_b}{\sqrt{\Delta g d_{50}}} \sim \lambda^{\frac{(1+\sigma)}{4}} f^{\frac{1}{2}}(D) \tag{2}$$

It is worth mentioning that for a hydraulically rough flow regime the function f(D) tends to a constant value.

Figure 6 depicts the vertical profiles of the time-averaged turbulent kinetic energy, K, normalized by $U_c{}^2$, at different downstream positions x/D from the bed sill. In the plane of flow symmetry ($y = 0$), $k = (\langle u'^2 \rangle + \langle w'^2 \rangle)/2$, where the angle brackets indicate the average over the length of the time series. In the energy inertial subrange, the energy cascade yields the '$\sigma = -5/3$' spectral law. Since the energy equilibrium in this energy subrange is maintained by the balance between the production and dissipation rate of the turbulent kinetic energy, the energy spectrum function is scaled as $E(\kappa) \sim \varepsilon^{2/3} \kappa^{-5/3}$, where ε is the turbulent kinetic energy dissipation rate and κ is the wavenumber. In this case, the turbulent kinetic energy can be scaled as $K \sim \int \varepsilon^{2/3} \kappa^{-5/3} d\kappa$. Figure 6 clearly highlights a spatial variation of K within the scour flow-field. In Figure 6, all the vertical profiles, except that at $x/x_s = 1.33$, show a decrease of K as going down towards the scour bed, but with different reduction rates that depend on the downstream positions x/x_s. At $x/x_s = 0.67$, K experiences both the maximum values and the maximum reduction rate. This implies that the jet-like region, indicated by region 1 in Figure 2, is a location of maximum turbulent energy production. At $x/x_s = 1.33$, K shows an increase with increasing depth, it attains a maximum value $O(3 \times 10^{-2})$ at $Z/z_s = -0.6$ and then begins to decrease going down to the bed-flow interface. This fact can be explained by the transition effect between regions 1 and 3 (see Figure 2), where the jet-like diffusion is accompanied by high levels of flow-turbulence intensities and large kinetic energy production, as also observed in a previous study by Ben Meftah and Mossa [20].

At $x/x_s > 1.33$, k undergoes a sharp decrease, showing values $O(10^{-2})$, it also shows a very gradual reduction going downwards. At equilibrium scour condition, k shows very small values, ranging between 0.005 and 0.013, near the bed-flow interface for the different downstream positions x/x_s. In the scour hole, at equilibrium condition, the stability of the sediment particles is due to the significant decrease of turbulent energy production at the bed-flow interface.

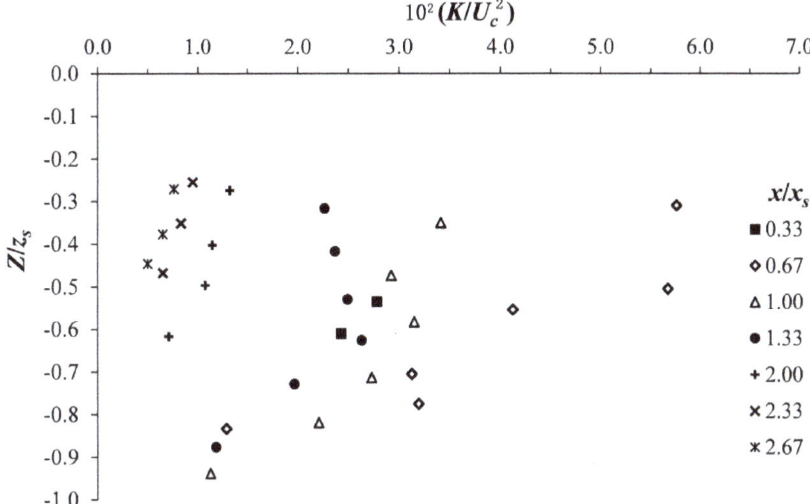

Figure 6. Vertical profiles of normalized turbulent kinetic energy, k/U_c^2, in scour hole and at different downstream positions x/x_s along the plane of flow symmetry ($y = 0$).

3.3. Turbulent Length Scales

Determination of the eddies scales in a turbulent flow is of crucial importance for experimental and numerical investigations, defining suitable domain-dimensions (area or volume) for computation [20]. Since the condition of incipient movement of the sediment particles is significantly influenced by the size of turbulent eddies, in this section, we try to experimentally determine the characteristic eddy length scales of the turbulent flow in the scour hole at equilibrium condition. The integral length scale L_i ($= L_x$, L_z) is simply calculated as the product of the integral time scale T_i and the local time-averaged velocity U_i ($= U$, W), where T_i ($= T_x$, T_z) is computed integrating the autocorrelation of the measured instantaneous flow velocity $u_i(t)$ [$= u(t)$, $w(t)$]. The autocorrelation of the measured instantaneous flow velocity was determined after a spectral analysis of the flow velocity fluctuation at the different measurement points.

Figures 7 and 8 depict the vertical profiles of the integral length scales L_x and L_z, respectively, normalized by the mean average diameter of sediments, d_{50}, at different downstream positions x/x_s. The processed data and the downstream positions x/x_s are the same as those covered in the previous sections. Herein, L_x is the integral length scale in the x-direction and L_z in the z-direction, calculated by means of the variables (U, T_x) and (W, T_z), respectively.

Contrary to what observed in Figures 3–6 for the flow turbulence properties in the scour hole, Figure 7 shows that larger values of L_x occur almost at the location of lower turbulence levels. At $x/x_s = 0.33$ and 0.64, i.e., the positions of the jet-like flow and jet diffusion in regions 1 and 2 (Figure 2), L_x has a size of the order of $1 \div 5d_{50}$. At $x/x_s = 1$, L_x is considerably increased, compared to the values at the upstream positions $x/x_s = 0.33$ and 0.64. It vertically decreases, going down towards the scour bed, from value $O(10d_{50})$ to value $O(d_{50})$ near the sediment bed. At $x/x_s = 1.33$, L_x shows a maximum value $O(27d_{50})$ and it monotonically decreases with increasing depth, reaching a value $O(7d_{50})$ close to the scour bed. At $x/x_s > 1.33$, a sharp increase of L_x can be clearly noted. It shows an average size of $O(40d_{50})$, and it ranges between a minimum and maximum of $20d_{50}$ and $54d_{50}$, respectively. Contrary to the region of high turbulence levels ($x/x_s < 1$), at $x/x_s > 1.33$, the L_x-magnitudes rapidly decrease with increasing flow depth, yielding maximum gradient values along the vertical. The integral length scale distribution in the scour hole at equilibrium provides an integrated hydrodynamic picture on the formation and relative macroscopic scales of the turbulent eddies.

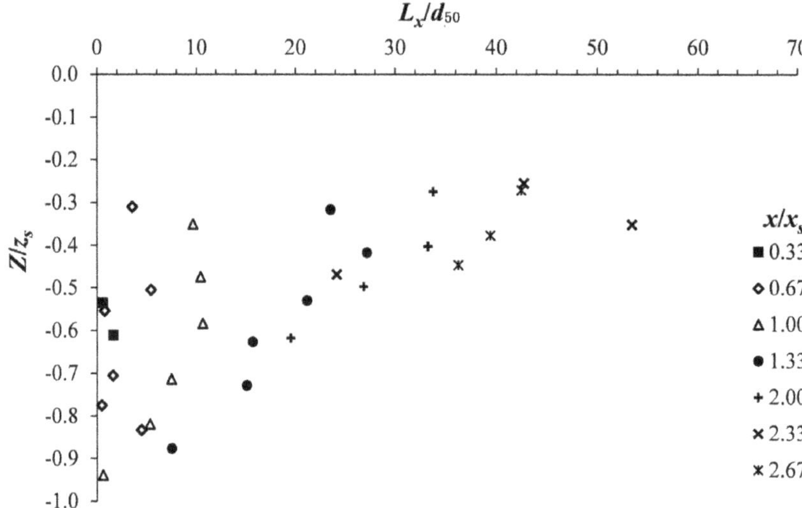

Figure 7. Vertical profiles of normalized turbulent length scales, L_x/d_{50}, at different downstream positions x/x_s at the plane of flow symmetry ($y = 0$).

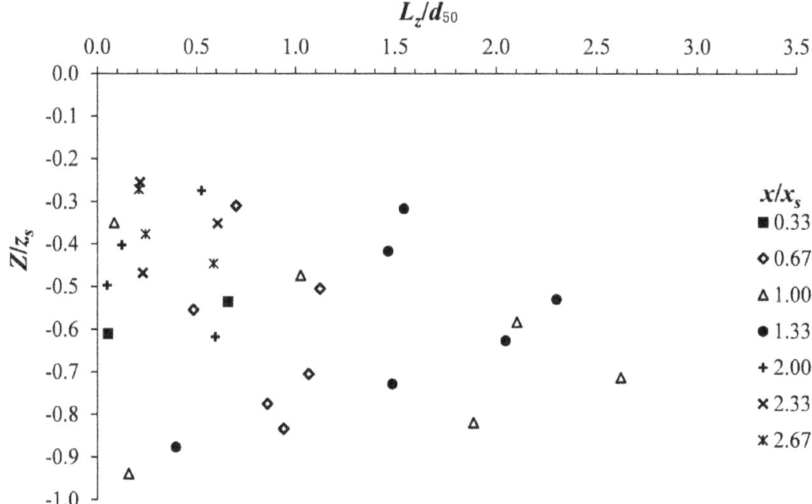

Figure 8. Vertical profiles of normalized turbulent length scales, L_z/d_{50}, at different downstream positions x/x_s at the plane of flow symmetry ($y = 0$).

Figure 8 illustrates the dimensionless integral length scale L_z/d_{50} versus the dimensionless vertical position Z/z_s in the scour hole at equilibrium phase. Contrary to what is shown in Figure 7 with L_x, Figure 8 indicates that the larger values of L_z appear at the location of higher turbulence levels. At $x/x_s \leq 1.33$, L_z shows both the largest values and the highest gradient along the vertical. It attains maximum values of almost $3d_{50}$ at $x/x_s = 1$ and 1.33. These values significantly decrease near the scour bed to an order of $0.1d_{50}$ to $0.4d_{50}$. This behavior could play an important role in increasing the vertical lifting force to move the sediment particles before reaching an equilibrium condition. At $x/x_s = 0.33$ and 1.33, L_z slightly decreases to values $O(1d_{50})$. This distribution of L_z at these positions seems reasonable, as it is strongly influenced by the incoming inclined flow-jet in the scour hole, which increases the vertical velocity component. At $x/x_s > 1.33$, L_z shows very small values, less than $1d_{50}$. This is explained by the smooth outflow from region 3, as observed in Figure 2, where a streamwise velocity dominance over the vertical component occurs.

The results obtained from the distribution of the integral length scales L_x and L_z seem to indicate that, in addition to the drag force, the erosion capacity of the flow increases with the increase of the vertical lifting force acting on the sediment particles. The vertical lifting force, essential for moving the sediment particles, is a direct result of an appropriate magnitude of the vertical velocity components.

Figure 9 displays the normalized Kolmogorov's micro-scale η/d_{50} versus the normalized vertical coordinate Z/z_s at different distances from the bed sills x/x_s. The main observation from Figure 9 is the expected significant reduction of η compared to L_x and L_z, by an average (over all measured points) factor of 340 and 25, respectively. Through the scour hole, η shows values ranging between $0.017d_{50}$ to $0.044d_{50}$, an equivalent of 0.04 to 0.15 mm. For the positions $x/x_s \leq 1.33$, η shows the smallest values, which are almost of constant magnitude along the vertical and are very comparable, indicating a kind of local isotropy of the flow turbulence at these scales. For $x/x_s > 1.33$, η shows an increase by an almost factor of 1.5 at $x/x_s = 2$, presenting an average length scale of the order of $0.56d_{50}$, and then continues to increase monotonically reaching an average value of the order of $0.73d_{50}$, quite constant at both positions $x/x_s = 2.33$ and 2.67. Figure 9 indicates that for all profiles, regardless of the positions x/x_s, η/d_{50} is invariant along the vertical direction. Figure 9 also points out that the size of η is influenced by the level of the flow turbulence intensity. In areas of high turbulence, such as at $x/x_s \leq 1.33$, η considerably decreases to smaller values. This may be explained by the diffusion of the inlet jet-like flow in the scour hole, which induces the formation of small eddy scales that increase velocity fluctuation.

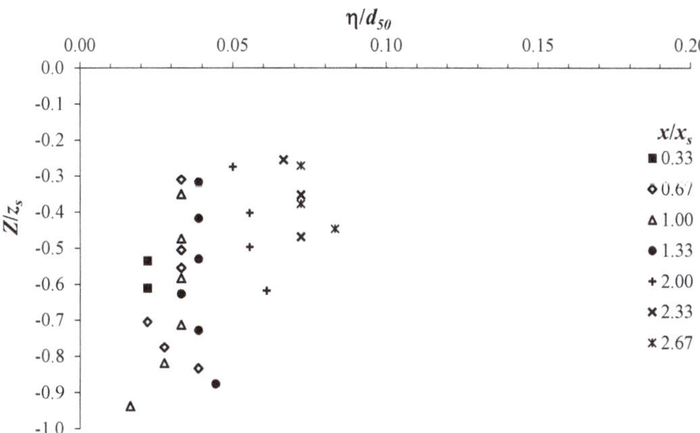

Figure 9. Vertical profiles of normalized Kolmogorov's micro-scale, η/d_{50}, at different downstream positions x/x_s at the plane of flow symmetry ($y = 0$).

3.4. Scaling of the Maximum Scour Depth at Equilibrium

In this section we focus on the scaling of the maximum scour depth at the equilibrium condition, derived from the phenomenological theory of turbulence. At equilibrium, the scour hole between two consecutive bed sills is shown schematically in Figure 10. The scour hole is caused by the effect of a sort of an entering jet flow of both initial thickness and initial velocity comparable, respectively, to the flow depth, h_c, and its corresponding velocity, U_c, over the sill. In Figure 10, q is the unit water discharge, l_s is the equilibrium scour length and h_t is the tailwater depth at the position x_s of the maximum scour depth at equilibrium, defined as the vertical distance between the initial bed profile and the free-surface flow (at x_s the total flow depth is $h_s = z_s + h_t$). The effective parameter of the scour hole, illustrated in Figure 10, can be expressed as follows:

$$h_s = z_s + h_t = f(h_c, U_c, d_{50}, L, S_0, g, \rho_w, \rho_s, \nu) \tag{3}$$

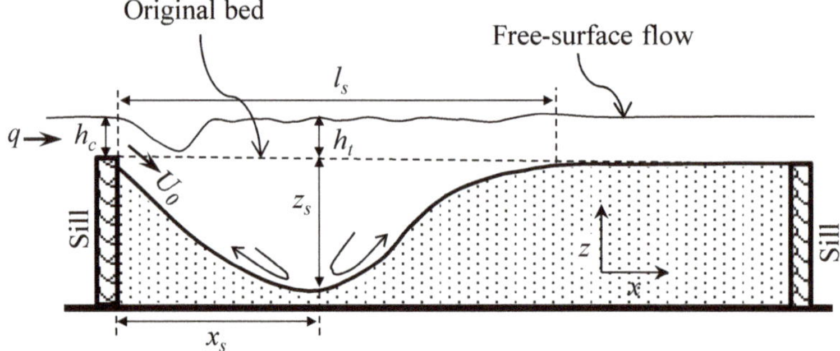

Figure 10. Definitional sketch of scour hole downstream of bed sills, Uo indicates the initial jet velocity.

The application of dimensional analysis to the variables of Equation (3) leads to the following expression:

$$\frac{h_s}{h_c} = \frac{z_s + h_t}{h_c} = f\left(\lambda_c, \frac{LS_0}{h_c}, F_{dc}, Re_c\right) \tag{4}$$

For fully turbulent flow, $Re_g > 70$ [31,32], the dependence upon the Reynolds number, Re_c, could be neglected, and therefore Equation (4) can be finally expressed as:

$$\frac{h_s}{h_c} = \frac{z_s + h_t}{h_c} = f\left(\lambda_c, \frac{LS_0}{h_c}, F_{dc}\right) \tag{5}$$

The application of the continuity equation between the section at $x = 0$ (over the bed sill) and the section at x_s, position of maximum scour, yields:

$$\begin{aligned} U_c h_c &= U_s h_s = U_s(z_s + h_t) \\ \frac{U_c}{U_s} &= \frac{h_s}{h_c} = \frac{(z_s + h_t)}{h_c} \end{aligned} \tag{6}$$

From Equation (2), we can obtain a scaling of the mean velocity U_s as:

$$U_s \propto U_b \sim \sqrt{\Delta g d_{50}} \lambda^{\frac{(1+\sigma)}{4}} f^{\frac{1}{2}}(D) = \sqrt{\Delta g d_{50}} \left(\frac{d_{50}}{h_s}\right)^{\frac{(1+\sigma)}{4}} f^{\frac{1}{2}}(D) \tag{7}$$

Substituting Equation (7) into Equation (6) produces:

$$\frac{h_s}{h_c} = \frac{z_s + h_t}{h_c} \propto F_{dc}^{\frac{4}{(3-\sigma)}} \left(\frac{d_{50}}{h_c}\right)^{-\frac{(1+\sigma)}{(3-\sigma)}} f^{-\frac{2}{(3-\sigma)}}(D) = F_{dc}^{\frac{4}{(3-\sigma)}} \lambda_c^{-\frac{(1+\sigma)}{(3-\sigma)}} f^{-\frac{2}{(3-\sigma)}}(D) \tag{8}$$

Since for rough turbulent flow $f(D)$ tends to be a constant function, Equation (8) can be reduced to the following form:

$$\frac{h_s}{h_c} = \frac{z_s + h_t}{h_c} \propto F_{dc}^{\frac{4}{(3-\sigma)}} \lambda_c^{-\frac{(1+\sigma)}{(3-\sigma)}} \tag{9}$$

It should be mentioned here that the scaling law of the maximum scour depth at equilibrium, investigated applying the phenomenological theory of turbulence, as shown by Equation (9), is explicitly expressed as a function of only the dimensionless parameters F_{dc} and λ, also appearing in Equation (5) through dimensional analysis. It should be noted that, contrary to dimensional analysis, the application of the phenomenological theory of turbulence establishes a unique and complete (with defined exponents) relationship between some of the characteristic variables of the scour hole. Since the flow is fully turbulent within the scour hole and the bed sediment is characterized by a relative roughness in the range of $10^{-4} < \lambda_c < 10^{-1}$ [30], the scaling law is validated for $\sigma = -5/3$, in the energy inertial subrange, and thus $F_{dc}^{(4/(3-\sigma))} \lambda_c^{(-(1+\sigma)/(3-\sigma))} = F_{dc}^{(6/7)} \lambda_c^{(1/7)}$. In order to experimentally evaluate the validity of this scaling law, shown in Equation (9), in Figure 11 we plot measured values of $(z_s + h_t)/h_c$ as a function of $F_{dc}^{(6/7)} \lambda_c^{(1/7)}$. The experimental data used for this scope are those illustrated in Table 1, acquired by Ben Meftah and Mossa [3].

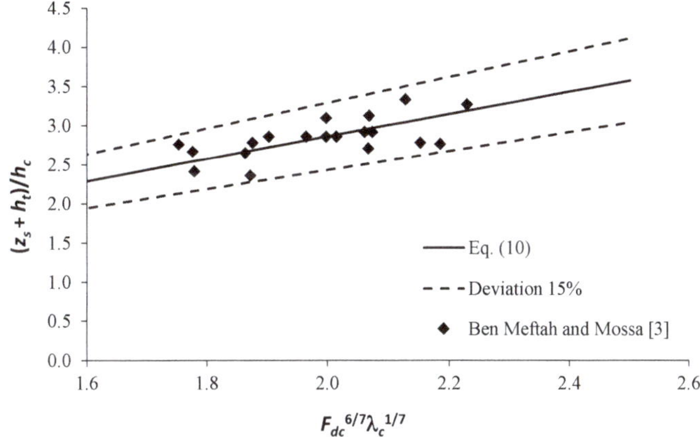

Figure 11. Validation of the scaling law of the maximum scour depth downstream of bed sills obtained by applying the fundamental laws of turbulent energy spectrum with $\sigma = -5/3$, and thus $F_{dc}^{(4/(3-\sigma))} \lambda_c^{(-(1+\sigma)/(3-\sigma))} = F_{dc}^{(6/7)} \lambda_c^{(1/7)}$. The data of Ben Meftah and Mossa [3] were obtained downstream of bed sills in alluvial channels and with a bed sediment size of $d_{50} = 1.8$ mm.

Figure 11 clearly shows that the data of the normalized maximum scour depth $(z_s + h_t)/h_c$ plotted versus the scaling law $F_{dc}^{(6/7)} \lambda_c^{(1/7)}$ tend to collapse into a single curve. From the interpolation of the data shown in Figure 11 the following expression, predicting the maximum scour depth at equilibrium, is obtained:

$$\frac{h_s}{h_c} = \frac{z_s + h_t}{h_c} = 1.43 F_{dc}^{\frac{6}{7}} \lambda_c^{\frac{1}{7}} \tag{10}$$

It is worth mentioning that Equation (10) is the regression fitting of data collected downstream of bed sills in alluvial channels. As reported in Ben Meftah and Mossa [3], the distance L between sills varies between 1 and 4 m and the non-cohesive bed sediment consists of very coarse sand particles of mean average size d_{50} = 1.8 mm. Despite the non-presence of the parameter LS_0/h_c in Equation (10), with respect to Equation (5), which takes into consideration the effects of both the distance and the slope between sills, Equation (10) suitably predicts the maximum scour depth downstream of bed sills at equilibrium condition. This is proved by the low deviation of the data, ±15%, from the best fit line of Equation (10).

To give more validity to this approach, in addition to the data obtained by Ben Meftah and Mossa [3], in Figure 12 we also plot data previously obtained by Bormann and Julien [4]. The scour data of Bormann and Julien [4] were collected downstream of a grade control structure of different face slopes (18°, 45°, and 90° with respect to the horizontal) in a sand-bed channel of d_{50} equal to 0.3 mm and 0.45 mm. Due to the complexity of the phenomenon and the wide variety of conditions (including different characteristics of the sediment particles, different geometries of the grade control structure, and different entering jet typologies), the data by Bormann and Julien [4], plotted according to the scaling law of Equation (9), show more scattering than those by Ben Meftah and Mossa [3]. As compared to previous studies [4,33], when data were fitted by other scaling laws, the scaling law of Equation (9) predicts with more accuracy the maximum equilibrium scour depth, as shown by the 50%-deviation from the best fit line of Equation (11). In fact, the normalized scour depth $(z_s + h_t)/h_c$ by Bormann and Julien [4] shows an overall increasing trend as a function of $F_{dc}{}^{(6/7)}\lambda_c{}^{(1/7)}$. A linear regression analysis of these data leads to the following expression for predicting the equilibrium scour depth at grade control structures (under Bormann and Julien's [4] conditions):

$$\frac{h_s}{h_c} = \frac{z_s + h_t}{h_c} = 0.53 F_{dc}{}^{\frac{6}{7}}\lambda_c{}^{\frac{1}{7}} \tag{11}$$

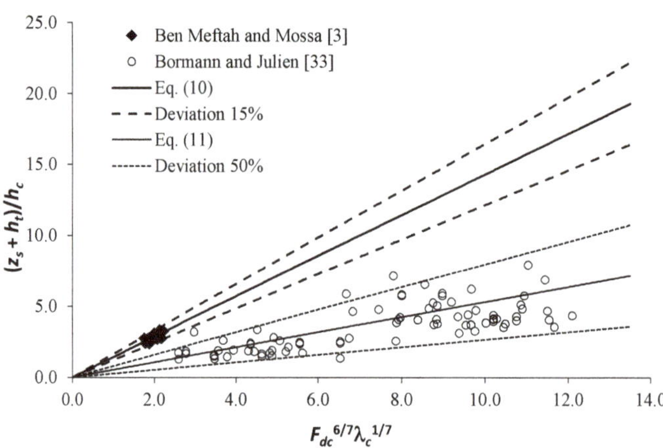

Figure 12. Validation of the scaling law of the maximum scour depth using data obtained by Bormann and Julien [4] downstream of grade control structures of different features and with sediment sizes of d_{50} ranging between 0.30 and 0.45 mm, Re_g ranging between 396 and 2093 and with relative roughness in the range of $10^{-4} < \lambda_c < 10^{-2}$. The scaling law is validated for $\sigma = -5/3$, in the energy inertial subrange.

It is worth noting that the scaling approach introduced by Equation (9) to predict the equilibrium scour depth is only expressed as a function of the densimetric Froude, F_{dc}, and the relative roughness, λ_c, with exponents function of σ. It does not involve all the characteristic parameters affecting the scour features, i.e., the parameter LS_0/h_c that appears in Equation (5) through dimensional analysis. In any case, its application shows reasonable accuracy in predicting the equilibrium scour depth in different hydraulic structures for sand-bed rivers. The limitation of this approach in the present form, however, is that it does not exhibit a general proportional relationship for all hydraulic structure typologies, as shown in Equations (10) and (11) by the coefficients 1.43 and 0.53. This implies that a further better combination between this scaling approach and other characteristic parameters, obtained through dimensional analysis, can lead to finding a general expression to predict the maximum equilibrium scour depth.

4. Conclusions

Considering that an exhaustive prediction of the scour development is fundamental for structural stability, we examined experimentally the local scouring processes downstream of hydraulic grade control structures. This study focuses on the analysis of flow hydrodynamic structures within the scour hole at equilibrium and on the application of the phenomenological theory of turbulence to find a scaling law predicting the equilibrium scour depth.

Specifically, by means of velocity measurements, we observed that at equilibrium conditions the flow in the scour hole is fundamentally characterized by three distinct regions: (i) A first region consists of a free entering jet flow, plunging from the crest of bed sill into the scour hole and strongly eroding the bed sediments; (ii) a second region located near the scour bottom, extending upstream due to vortex formations (eddies) generated by the jet diffusion, allowing to reach the equilibrium condition; and (iii) a third less-turbulent region, localized downstream of the first and the second regions and characterized by an almost unidirectional flow in the x-direction.

The detailed analysis of the flow turbulence characteristics in the equilibrium scour hole showed that both the Reynolds shear stresses and turbulence intensities (the streamwise and the vertical ones) exhibit their greatest values in the first region, where the approaching jet has effects, while a sharp reduction of them occurs in the third region. Moreover, these turbulent properties (Reynolds shear stresses and turbulence intensities) show negative vertical gradient going down towards the scour bed. A heterogeneous spatial distribution of these turbulent features is especially evident in the first and second regions, where the jet effect is quite strong. The analysis of the turbulent kinetic energy shows that the region of jet-like flow, region 1, is a location of maximum turbulent energy production. At equilibrium condition, the stability of sediment particles is due to a significant decrease of turbulent energy production at the bed-flow interface.

The distribution of the longitudinal and vertical integral length scales, in the plane of flow symmetry, was also analyzed. The larger values of L_x occur almost at the location of lower turbulence levels. On the contrary, L_z shows maximum values at the regions of higher turbulence levels. This distribution of L_x and L_z reflects the role of the flow turbulent eddies in the incipient motion of the sediment particles. The increase of L_z, related to the increase of the vertical velocity component, enhances the vertical lifting force, giving more possibilities for particles to move.

Finally, the phenomenological theory of turbulence was applied to the scour hole, deducing a new scaling law of the maximum scour depth at equilibrium in non-cohesive bed rivers. Once fixed the spectral exponent of the turbulent energy spectrum, the scaling low becomes a simple function of the densimetric Froude number, F_{d_c}, and the relative roughness, λ_c. This scaling law was evaluated and validated using laboratory measured scouring data of sand-bed with a relative roughness in the range of $10^{-4} < \lambda_c < 10^{-1}$. It showed a quite reasonable accuracy in predicting the equilibrium scour depth in different hydraulic structures.

Author Contributions: M.B.M. performed the experiments, analyzed the data, designed the study and wrote the paper; F.D.S., D.D.P. and M.M. contributed suggestions, discussions and reviewed the manuscript. All authors have read and agreed to the published version of the manuscript.

Funding: This research recevied no external funding.

Acknowledgments: The experiments were carried out at the Hydraulic Laboratory of the Mediterranean Agronomic Institute of Bari (Italy).

Conflicts of Interest: The authors declare that they have no conflicts of interest.

References

1. Adduce, C.; Sciortino, G. Scour due to a horizontal turbulent jet: Numerical and experimental investigation. *J. Hydraul. Res.* **2006**, *44*, 663–673. [CrossRef]
2. Balachandar, R.; Kells, J.A. Local channel in scour in uniformly graded sediments: The time-scale problem. *Can. J. Civ. Eng.* **1997**, *24*, 799–807. [CrossRef]
3. Ben, M.M.; Mossa, M. Scour holes downstream of bed sills in low-gradient channels. *J. Hydraul. Res.* **2006**, *44*, 497–509.
4. Bormann, N.E.; Julien, P.Y. Scour downstream of grade-control structures. *J. Hydraul. Eng.* **1991**, *117*, 579–594. [CrossRef]
5. Carstens, M.R. Similarity laws for localized scour. *J. Hydraul. Div.* **1966**, *92*, 13–36.
6. D'Agostino, V.; Ferro, V. Scour on alluvial bed downstream of grade-control structures. *J. Hydraul. Eng.* **2004**, *130*, 24–37. [CrossRef]
7. Espa, P.; Sibilla, S. Experimental study of the scour regimes downstream of an apron for intermediate tailwater depths. *J. Appl. Fluid Mech.* **2014**, *7*, 611–624.
8. Gaudio, R.; Marion, A.; Bovolin, V. Morphological effects of bed sills in degrading rivers. *J. Hydrau. Res.* **2000**, *38*, 89–96. [CrossRef]
9. Lenzi, M.A.; Marion, A.; Comiti, F. Local scouring at grade-control structures in alluvial mountain rivers. *Water Resour. Res.* **2003**, *39*, 1176. [CrossRef]
10. Mason, P.J.; Arumugam, K. Free jet scour below dams and flip buckets. *J. Hydraul. Eng.* **1985**, *111*, 220–235. [CrossRef]
11. Pagliara, S.; Amidei, M.; Hager, W.H. Hydraulics of 3D plunge pool scour. *J. Hydraul. Eng.* **2008**, *134*, 1275–1284. [CrossRef]
12. Papanicolaou, A.N.; Bressan, F.; Fox, J.; Kramer, C.; Kjos, L. Role of structure submergence on scour evolution in gravel bed rivers: Application to slope-crested structures. *J. Hydraul. Eng.* **2018**, *144*, 1087–1093. [CrossRef]
13. Wang, L.; Melville, B.W.; Guan, D.; Whittaker, C.N. Local scour at downstream sloped submerged weirs. *J. Hydraul. Eng.* **2018**, *144*, 04018044. [CrossRef]
14. Tregnaghi, M.; Marion, A.; Gaudio, R. Affinity and similarity of local scour holes at bed sills. *Water Resour. Res.* **2007**, *43*, W11417. [CrossRef]
15. Lu, J.Y.; Hong, J.H.; Chang, K.P.; Lu, T.F. Evolution of scouring process downstream of grade-control structures under steady and unsteady flows. *Hydrol. Process.* **2013**, *27*, 2699–2709. [CrossRef]
16. Manes, C.; Brocchini, M. Local scour around structures and the phenomenology of turbulence. *J. Fluid Mech.* **2015**, *779*, 309–324. [CrossRef]
17. Gioia, G.; Bombardelli, F.A. Localized turbulent flows on scouring granular beds. *Phys. Rev. Lett.* **2005**, *95*, 014501. [CrossRef]
18. Bombardelli, F.A.; Gioia, G. Scouring of granular beds by jet-driven axisymmetric turbulent cauldrons. *Phys. Fluids* **2006**, *18*, 088101. [CrossRef]
19. Ali, S.Z.; Dey, S. Impact of phenomenological theory of turbulence on pragmatic approach to fluvial hydraulics. *Phys. Fluids.* **2018**, *30*, 045105. [CrossRef]
20. Meftah, M.B.; Mossa, M. Turbulence measurement of vertical dense jets in crossflow. *Water* **2018**, *10*, 286. [CrossRef]
21. Meftah, M.B.; Malcangio, D.; De Serio, F.; Mossa, M. Vertical dense jet in flowing current. *Environ. Fluid Mech.* **2018**, *18*, 75–96. [CrossRef]
22. Meftah, M.B.; De Serio, F.; Mossa, M.; Pollio, A. Experimental study of recirculating flows generated by lateral shock waves in very large channels. *Environ. Fluid Mech.* **2008**, *8*, 215–238. [CrossRef]
23. Meftah, M.B.; Mossa, M. Partially obstructed channel: Contraction ratio effect on the flow hydrodynamic structure and prediction of the transversal mean velocity profile. *J. Hydrol.* **2016**, *542*, 87–100. [CrossRef]
24. Meftah, M.B.; Mossa, M. A modified log-law of flow velocity distribution in partly obstructed open channels. *Environ. Fluid Mech.* **2016**, *16*, 453–479. [CrossRef]
25. Ben Meftah, M.; De Serio, F.; Mossa, M. Hydrodynamic behavior in the outer shear layer of partly obstructed open channels. *Phys. Fluids.* **2014**, *26*, 65102. [CrossRef]
26. Ben Meftah, M.; Mossa, M. Prediction of channel flow characteristics through square arrays of emergent cylinders. *Phys. Fluids.* **2013**, *25*, 45102. [CrossRef]

27. Ben Meftah, M.; Mossa, M.; Pollio, A. Considerations on shock wave/boundary layer interaction in undular hydraulic jumps in horizontal channels with a very high aspect ratio. *Eur. J. Mech. B Fluids.* **2010**, *29*, 415–429. [CrossRef]

28. Scurlock, S.M.; Thornton, C.I.; Abt, S.R. Equilibrium scour downstream of three-dimensional grade control structures. *J. Hydraul. Eng.* **2011**, *138*, 167–176. [CrossRef]

29. Ghodsian, M.; Mehraein, M.; Ranjbar, H.R. Local scour due to free fall jets in non-uniform sediment. *Sci. Iranica.* **2012**, *19*, 1437–1444. [CrossRef]

30. Ali, S.Z.; Dey, S. Origin of the scaling laws of sediment transaport. *Proc. R. Soc. A* **2017**, *473*, 20160785. [CrossRef]

31. De Vincenzo, A.; Brancati, F.; Pannone, M. An experimental analysis of bed load transport in gravel-bed braided rivers with high grain Reynolds numbers. *Adv. Water Res.* **2016**, *94*, 160–173.

32. Mirauda, D.; De Vincenzo, A.; Pannone, M. Statistical characterization of flow field structure in evolving braided gravel beds. *Spat. Stat.* **2019**, *34*, 100268. [CrossRef]

33. Meftah, M.B.; Mossa, M. New approach to predicting local scour downstream of grade-control structure. *J. Hydraul. Eng.* **2019**, *146*, 04019058. [CrossRef]

Article

Flow–Sediment Turbulent Ejections: Interaction between Surface and Subsurface Flow in Gravel-Bed Contaminated by Fine Sediment

Bustamante-Penagos N. [1,*] and Niño Y. [1,2]

[1] Department of Civil Engineering, Faculty of Physical and Mathematical Sciences, Universidad de Chile, Santiago 8370449, Chile; ynino@ing.uchile.cl

[2] Advanced Mining Technology Center, Faculty of Physical and Mathematical Sciences, Universidad de Chile, Santiago 8370449, Chile

* Correspondence: nataliabustamante@ug.uchile.cl

Received: 3 March 2020; Accepted: 23 April 2020; Published: 3 June 2020

Abstract: Several researchers have studied turbulent structures, such as ejections, sweeps, and outwards and inwards interactions in flumes, where the streamwise velocity dominates over vertical and transversal velocities. However, this research presents an experimental study in which there are ejections associated with the interchange between surface and subsurface water, where the vertical velocity dominates over the streamwise component. The experiment is related to a surface alluvial stream that is polluted with fine sediment, which is percolated into the bed. The subsurface flow is modified by a lower permeability associated with the fine sediment and emerges to the surface current. Quasi-steady ejections are produced that drag fine sediment into the surface flow. Particle image velocimetry (PIV) measured the velocity field before and after the ejection. The velocity data were analyzed by scatter plots, power spectra, and wavelet analysis of turbulent fluctuations, finding changes in the distribution of turbulence interactions with and without the presence of fine deposits. The flow sediment ejection changes the patterns of turbulent structures and the distribution of the turbulence interactions that have been reported in open channels without subsurface flows.

Keywords: ejections; turbulence interactions; gravel beds; sediment transport; surface and subsurface flows

---

1. Introduction

Landslides, volcanos, or anthropogenic changes may modify the availability of fine sediment in rivers, reservoirs, or lakes [1]. Fine sediment can cause pollution in gravel beds. This contamination has a high environmental impact because the porosity of the gravel beds is a reservoir for the deposition of fine sediment [2]. The intrusion of fine sediment into gravel bed streams generates changes in the hyporheic exchange, nutrients cycling, low oxygenation of fish eggs, etc. [3–5]. Additionally, the hyporreic zone also have an important coupling between the subsurface groundwater system and surface water, such as rivers or lakes and floodplains. This exchange is through the porous sediment, and it is characterized by the circulation of surface water into the alluvium and back to the river bed [6].

Moreover, fine sediments can move in suspension when the turbulent eddies have upward velocity components exceeding the fall velocity of fine sediments [7]. An increase in grain roughness can generate an increase in the vertical intensity of turbulence [8]. Moreover, strong upward turbulent ejections could provide the vertical anisotropy needed for suspension transport and the entrainment of the fine sediment [8–10]. In addition, the turbulent structures in smooth flumes have been studied by [11–14] and others, showing coherent structures such as individual hairpin vortices, which have a small scale motion $l \sim h$, hairpin vortices that have a large scale motion $l \sim 3h$, and super streamwise vortices, which have a super scale motion $l > 10h$, where l is the streamwise scale of the vortices and h is the water depth. However, ejections in a rough-bed are smaller near the bed and the secondary currents can change the ejection patterns [15]. Furthermore, several researchers such as [12,16–19] have investigated turbulence characteristics considering a Cartesian plane with streamwise velocity fluctuations, u', and vertical velocity fluctuations, w', where the interaction in the second and fourth quadrant (ejections $u' < 0$ & $w' > 0$ and sweeps $u' > 0$ & $w' < 0$), respectively, are more frequent than the interactions in the first and third quadrants (outward interaction $u' > 0$ & $w' > 0$ and inward interaction $u' < 0$ & $w' < 0$), respectively, because of the mean shear stress is positive (i.e., $\overline{u'w'} < 0$). Experimentally, researchers have used acoustic Doppler velocimetry (ADV), particle image velocimetry (PIV), laser Doppler velocimetry (LDA), or ultrasonic Doppler velocimeter (UDV) for the acquisition of velocity data. Niño and Musalem [16] used ADV for characterizing the turbulence interactions in a sand bed, reporting ejections and sweeps as the most frequent turbulence interactions. Niño and Musalem [16] also reported that the sweeps are more efficient for the entrainment of the particles into suspension than ejections. Sambrook Smith and Nicholas [17], Cooper et al. [18], and Chen et al. [20] implemented PIV to characterize the flow field and the turbulence properties. Sambrook Smith and Nicholas [17] experimented with the deposition of sand on gravel beds. They reported high velocities and high shear stresses that occur at the level of the crest of the major roughness elements and in their lee side; in addition, interactions such as sweeps and ejections are less frequent when roughness decreases, i.e., the main effect of fine sediment deposited in a gravel bed is to reduce the vertical velocities' gradients and shear stresses near the bed over the sand on the gravel bed.

The roughness and the flow depth can modify the turbulent structures at shallow flow conditions [21]. Roussinova [21] compared results for rough and smooth walls and found that the magnitude of the turbulence quantities are higher in the case of the rough wall, ejection events are prevalent over sweeps and in smooth wall cases, and there are ejections and sweeps in the vicinity of the hairpin vortex.

Manes et al. [22] compared the turbulence structure for permeable and impermeable beds in open channels, finding that large scale eddies generated within the surface flow have influence in the subsurface flow, and they think that it must be associated with pressure fluctuations.

Fourier series are often used to discuss the properties of a turbulent flow field [23,24]. The frequency analysis, for example, is derived from the Fourier spectrum. However, this analysis is for stationary signals and the Fourier transformation has no localization property, i.e., if a signal changes at one position, then the transform changes without the position of the change could be recognized "at a glance" [25,26]. For unsteady signals that have finite duration, such as in geophysical processes and hydrology, the wavelet transform and the cross wavelet transform are excellent tools for analyzing the physical relationships between the time series [27,28]. However, the open channel turbulent velocity fluctuations have been analyzed by Chen et al. [20] considering the wavelet coherency, i.e., measuring the wavelet correlation between two velocity series at a frequency f on a scale from zero to unity, finding that the wavelet analysis identified the scale of motions and the time of its occurrence.

The present study characterizes the turbulence structures associated with the interchange between subsurface and surface flows due to fine sediment, pumicite, deposited into the interstitial space, the pores of a gravel bed. Particle image velocimetry (PIV) was employed to measure the velocity field before and after the fine sediment was deposited. The velocity data were analyzed by a scatter plot of turbulent fluctuation, Fourier, and wavelet analysis, finding changes in the distribution of turbulence interactions for a flume with and without fine sediment deposits.

2. Materials and Methods

Experiments were carried out in an open channel with a sediment bed. With this experiment setup, it was possible to measure independently the surface and subsurface flow. In this research, an immobile bed was considered. The fine material with which the bed is polluted is pumicite, or natural pozzolan, a raw material of minute grains of volcanic glass and ashes with the characteristics of clay.

2.1. Experimental Set Up

The experimental facility is an open channel, 0.03 m wide, 0.58 m long, and 0.63 m deep. The facility is divided into three parts. Upstream are the surface and subsurface input flows and the location of the seeding particles for PIV. At the center of the structure are the open channel, the bed, the point where the pumicite mixture is spilled, and the PIV measurement area. Downstream are the surface and subsurface output discharges (Figure 1a). The sediment bed has two layers of sediment. The surface layer is of gravel, 20 mm thick and median diameter $D_g = 10$ mm, and the subsurface layer is of fine gravel, 390 mm thick, and mean diameter of $D_s = 2.45$ mm (Figure 1b). The grain size distributions of sediment are shown in Figure 2. The density of both materials, gravel and fine gravel, is 2.65 g/cm^3, whereas the pumicite has a characteristic diameter (D_{50}) of $D_c = 0.12$ mm (Figure 2) and a density of 1.7 g/cm^3. The pumicite was fed through an acrylic cone in the free surface, of 1.0 cm of diameter, during 6 s. The net weight of pumicite was 284.2 g for the experiments. Such feeding can simulate a soil failure that falls into the river. Furthermore, the pumicite is poured with a concentration of 57% by weight in water.

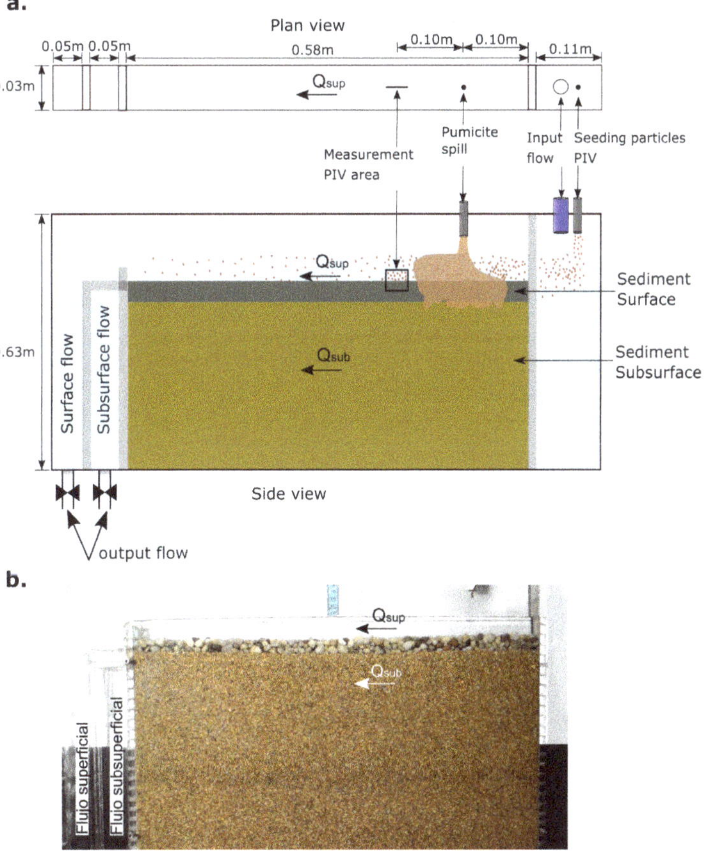

Figure 1. (**a**) experimental scheme used in the study; (**b**) experimental flume. The sediment column is used to measure high percorlations of fine sediment (pumicite) and subsurface flow. The surface flow rate is Q_{sur} and subsurface flow is Q_{sub}.

The measurement of flow rates was through volumetric gauges for the surface, Q_{sur}, and subsurface, Q_{sub}, flows in the experiments. A Photron FASTCAM Mini UX50 camera (San Diego, CA, USA) was used for Particle Image Velocimetry. This camera takes up to 2500 fps. A Nikon D3200 camera (Tokyo, Japan) was used for percolation analysis of pumicite. The Malvern Master Sizer 2000 equipment (Malvern, UK) of the Laboratory of Sedimentology of Universidad de Chile was used for measurement of the grain size distribution of pumicite. Experiments with several flow rates Q_{sur} and Q_{sub} are presented in Bustamante and Niño [29,30] for percolation of a variety of fine sediments, among them with pumicite. The experiments in this article are two, for the flow rates $Q_{sur} = 0.088$ L/s and $Q_{sub} = 0.008$ L/s. One with uniform flow with a friction slope $S_f = 0.017$, and water depth $H = 67$ mm, without fine sediment (E0) and the other of the same hydrodynamics conditions but with the spill of the fine sediment (E1). The article focuses, in this way, on the interaction between subsurface and surface flows due to the deposition of fine sediment in the bed by measuring turbulence in the surface flow.

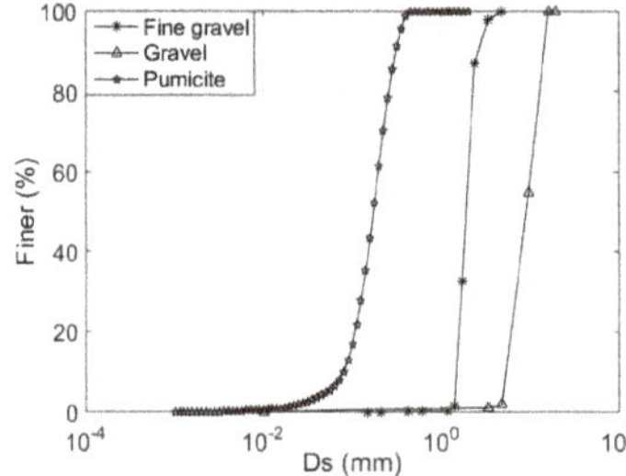

Figure 2. The grain size distribution of sediments used in the study.

2.2. Velocities

PIV was implemented to measure the flow field. The PIV particle tracers were of diameter 0.12 mm and density 1.7 g/cm^3. The flow was illuminated with a lamp with power of 18 W. The particle tracers are poured dry in low quantity in the stilling tank. Furthermore, for these conditions, the PIV particles are a good tracer because these particles only can move in suspension. Image acquisition was with the Fastcam Mini UX50 camera. It was implemented taking 250 fps during 140 s, before and after the spill of pumicite mixture, in the center of the cross section.

The velocity data were analyzed considering three methodologies. The first methodology is the estimation of the mean shear stress profile with the fluctuations of velocities with the spatially averaged open channel flow methodology proposed by Nikora et al. [31]. It was analyzed defining the total shear stress such as:

$$\tau_{tot} = \mu(\partial\overline{u})/\partial z - \rho\left\langle\overline{u'w'}\right\rangle - \rho\left\langle\tilde{u}\tilde{w}\right\rangle, \tag{1}$$

i.e., the mean shear stress has three components which are: viscous ($\mu(\partial\overline{u})/\partial z$), turbulent ($-\rho\left\langle\overline{u'w'}\right\rangle$), and form-induced ($-\rho\left\langle\tilde{u}\tilde{w}\right\rangle$), where μ is the kinematic viscosity, $\overline{u}$ is the mean velocity, u' and w' are the velocity fluctuations in x (streamwise) and z (vertical) components, respectively. $\tilde{u}$ and $\tilde{w}$ are the form-induced disturbance in the flow variables (where $\tilde{u} = \overline{u} - \langle\overline{u}\rangle$ and $\tilde{w} = \overline{w} - \langle\overline{w}\rangle$), x is the streamwise coordinate and z is the vertical coordinate. In addition, the spatial average mean velocity, $<\overline{u}>(z)$, results from this methodology. In Equation (1), the overbar means temporal average and the angular brackets mean spatial averages.

The second methodology considers the scatter plots of velocity fluctuations u' and w', which were analyzed considering that they could be represented by an ellipse, such as Equation (2), where α, Ra and Rb are the angle of rotation, major axis and minor axis, respectively [32]. Parameter α is obtained as the slope of a linear fit for the scatter plot between u' and w':

$$\frac{(x \cdot cos(\alpha) + y \cdot sin(\alpha))^2}{Ra^2} + \frac{(x \cdot sin(\alpha) + y \cdot cos(\alpha))^2}{Rb^2} = 1 \qquad (2)$$

Additionally, Niño and Musalem [16] and Wallace [19] measured the intensity of the ejections and sweeps in a measurement at one point by the parameter $K = u'w' / < u'w' >$. In this research, we have considered the parameter as: $K' = -K = -u'w' / | u'w' |$. The percentage of distribution of the K' parameter in an area of 73 mm × 68 mm at one point x and z of the experiment E0 is presented in Figure 3, where K'_{90} and K'_{10} are associated with the 10th and 90th percentiles of K', respectively. After finding the pairs (u', w') associated with K'_{90} and K'_{10}, the Ra and Rb can be defined. The methodology used for characterizing the ellipses is presented in Figure 4.

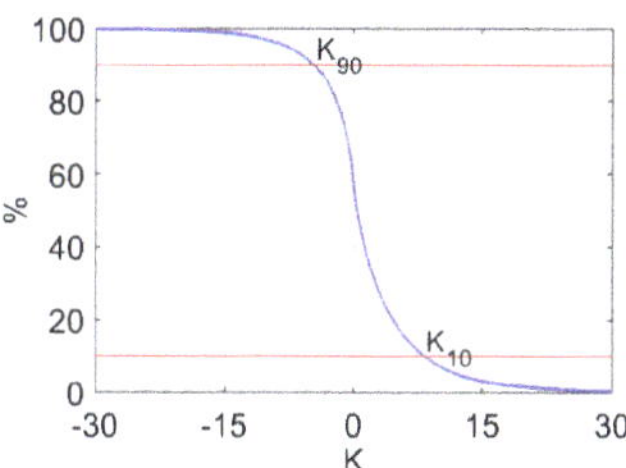

Figure 3. Percentage of distribution of parameter K.

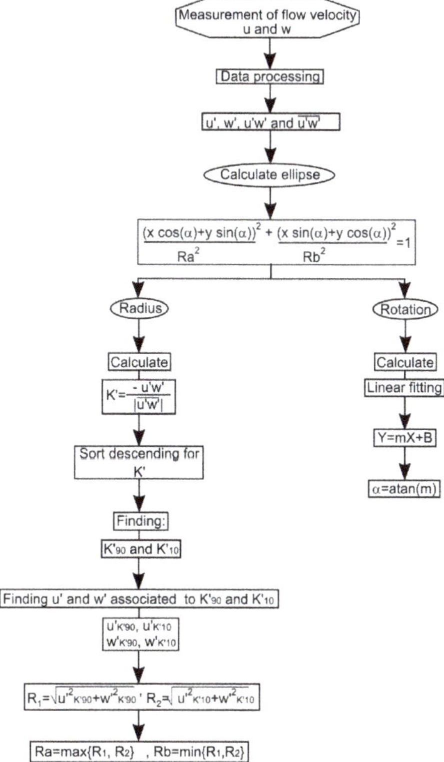

Figure 4. Methodology to characterize the scatter plot of velocity fluctuations through the ellipse parameters.

The third methodology was Fourier and wavelet analysis. The wavelet analysis has been implemented by [20,26,28,33,34] and others, in order to analyze different physical phenomena. In this research, the Morlet function was used as the mother wavelet. Morlet wavelet mother is a powerful tool for data analysis of low-oscillation functions [26].

3. Results

Hydraulic parameters for the experiment setup are the Froude number, $Fr = U/\sqrt{gH}$, Reynolds number, $Re = UH/\nu$, B/H is the aspect ratio between wide flume (B) and depth water (H), H/D_g is the ratio of depth water and gravel diameter, S_f is the friction slope (bed slope), and $U = Q_{sur}/(BH)$, as Table 1 shows.

Table 1. Hydraulic parameters.

Q_{sur}	Q_{sub}	U	Fr	Re	H/B	H/D_g	S_f
L/s	L/s	m/s	–	–	–	–	–
0.088	0.008	0.044	0.054	2933	2.2	6.7	0.017

In experiment E1, the fine sediment in the channel was poured locally, at the beginning of the flume ($x = 0.05$ m), as a hyperconcentrated mixture of fine sediment with water. The dynamics of this mixture considers suspension, deposition, and percolation. After deposition, the fine sediment changes the permeability of the gravel bed. Thus, the interaction between surface and subsurface flow generates ejections of water from the bed to the water column. These jets eject subsurface water with fine sediment of low density deposited in the interstices of the gravel bed. The length of the jets has been approximately 15 mm, as shown in Figure 5a.

These sediment ejections generate a low entrainment of fine sediment into suspension transport or bedload transport, i.e., the sediment ejected is deposited in the neighborhood of the ejection, forming craters as bedforms (Figure 5c). Then, the experiment E1 is to characterize the ejection of the subsurface flow, which is modified by a lower permeability associated with the fine sediment and emerges to the surface current.

Figure 5 is localized in the tank at 20 cm downstream of the center part of the facility as shown in Figure 1. The velocity data are taken at the center of the cross section and at the point where the sediment ejections appear. Furthermore, the velocity fluctuations and wavelet analysis are analyzed at 1.5 cm up from the of pumicite level. The velocity profiles were analyzed considering spatially averaged open-channel flow proposed by Nikora et al. [35].

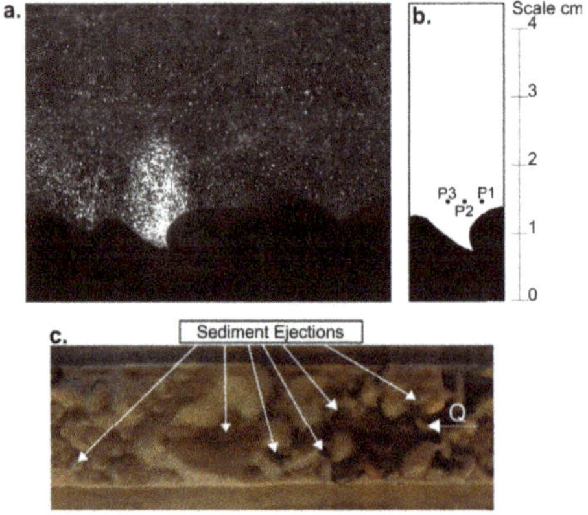

Figure 5. (**a**) sediment ejection for Q_{sur} = 0.088 L/s y Q_{sub} = 0.08 L/s; (**b**) measurement points; (**c**) top view of the sediment ejections.

3.1. Velocities and Shear Velocities

The velocity data were measured with PIV and processed with PIVLab. The spatial average mean velocity profile and the spatial average mean shear rate were obtained according to the double-averaged methodology proposed by Nikora et al. [35]. The spatial average mean velocity profile made dimensionless with the shear velocity for the experiments E0 and E1 are shown in Figure 6. Table 2 shows the shear velocities in both experiments, u_{*E0} and u_{*E1}. Shear velocities were calculated as $u_{*Ei} = \sqrt{\tau_{0Ei}/\rho}$, where τ_{0Ei} is the experimental shear stress in the bed and $i = 0, 1$, for experiment E0 or E1, respectively.

The dimensionless velocity profile for E1 is much more intense than for E0. This is because the shear velocity u_{*E0} is 39% higher than the shear velocity u_{*E1}. However, the ratio of the depth average mean velocities in E1 and E0, U_{E1}/U_{E0}, is 0.90. Therefore, the surface flow is faster in E0 than in E1 because the sediments ejection has a dominant vertical velocity. However, in an area without sediment ejection, the streamwise velocity in E1 has to be greater than E0 because there is outflow of the subsurface flow to the surface flow.

Table 2. Velocities and shear velocities.

U_{E0}	U_{E1}	u_{*E0}	u_{*E1}
m/s	m/s	m/s	m/s
0.058	0.052	0.0032	0.0023

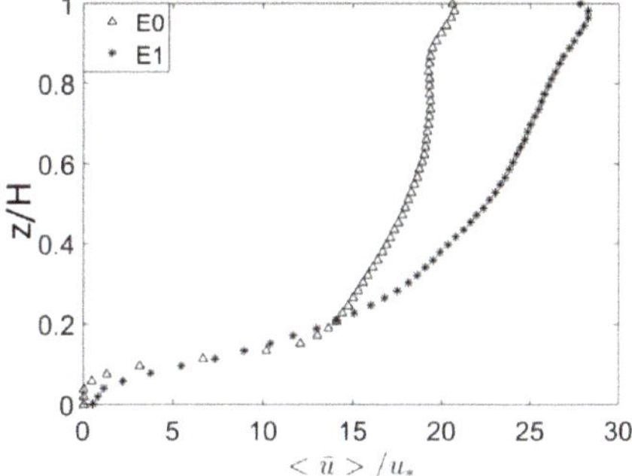

Figure 6. Velocity profiles double-averaged, for both experiments E0 and E1, before and after the spill of pumicite mixture, respectively.

In Figure 5b are the points that were analyzed for the turbulence interactions in one sediment ejection—that is, upstream of the center of ejection (P1), the center of ejection (P2) and downstream of the center of ejection (P3). Turbulence fluctuations were analyzed under three approaches: scatter plot of u' and w', velocity field, and wavelet analysis.

3.2. Velocity Field

Velocity vectors in the streamwise and vertical plane obtained through PVI processing are presented in Figure 7, for three different times, 40.00 s, 40.63 s, and 41.26 s. Figure 7 also presents the contour plot of the vertical velocity. The red polygon in that figure is limited where the vertical velocities are higher than the streamwise velocities in the ejection. Additionally, in Figure 7c, the blue lines show a coherent structure external to the movement we are observing, which is also a coherent structure from upstream. The vertical upward movement from the bed to the water column is associated with turbulent interactions of the ejection type. However, the sediment ejections reported in this research are associated with jets with sediment, due to the interaction between surface and subsurface flow, and this is different from the turbulent interactions of the ejection type investigated for rigid wall, as reported by [9,12,18,36,37] and others, who have analyzed turbulence interactions near the wall and have identified two main interactions as ejections and sweeps.

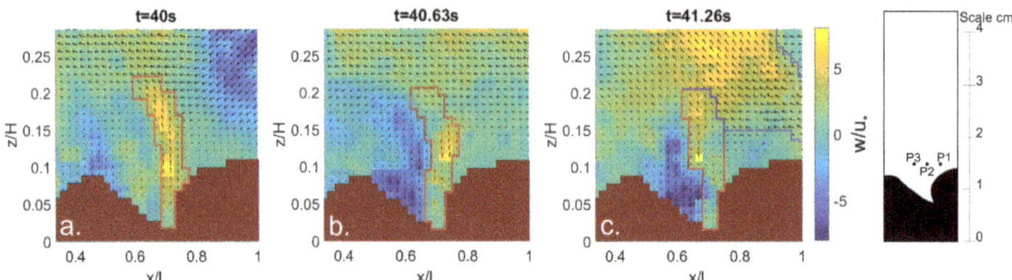

Figure 7. Velocity field for dimensionless vertical component (W/u*) and vectors of streamwise velocity and vertical velocity, measured experimentally for E1, at **a** t = 40 s, **b** t = 40.63 s, **c** t = 41.26 s. Measured field was Δx = L = 2.9 cm y Δz = 3.5 cm and flow depth H = 6.7 cm. Direction of the flow: from right to left.

In E1, the vertical velocities, w, made dimensionless with mean streamwise velocity, U, were analyzed before and after the spill of pumicite mixture as a function of dimensionless time tU/H. In this case, the vertical velocity time series at the water depth for the three positions into the sediment ejection, i.e., in each position, P1, P2, and P3, shown in Figure 5b, the velocities series were taken in the entire water column. However, for the case of E0, with no-spill of pumicite mixture, the vertical velocities are those associated only to P1 and P2. The vertical in P1 is a measure of the vertical velocity at a point on the gravel ridges (Figures 7 and 8a), while the vertical in P2 is a measure in the gravel pores (Figures 7 and 8b). The vertical velocities in E0, for $z/H > 0.1$ and in positions P1 and P2 follow Taylor's frozen turbulence hypothesis in the streamwise direction quite well (Figure 8a,b). Vertical velocities are higher for $z/H > 0.5 - 0.6$ than near the bed at P1 or P2 (Figure 8a,b). For position P2 near the bed, for $z/H < 0.1$, there is an area with high vertical velocity for $11 < tU/H < 61$ (Figure 8b) that is not detected for position P1. That is, the irregularities presented by the bed of gravel, with the ridges and low points, make Taylor's hypothesis of frozen turbulence invalid near the bottom.

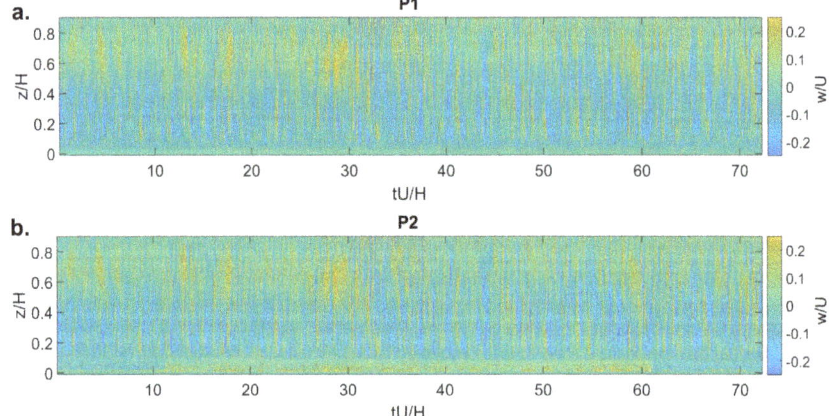

Figure 8. Vertical velocity made dimensionless with U, w/U in the experiment E0, with no-spill of pumicite mixture, for: (**a**) upstream of the center of the ejection, P1, (**b**) center of the ejection, P2.

Conversely, in E1, after the spill of pumicite mixture, it is found that the ejections generate streamwise changes in the vertical velocities. At P2, in the gravel pore, an ejection take place, with high positive vertical velocities, for $z/H < 0.2$ and $0 < tU/H < 80$ (Figure 9b). At P3, negative vertical velocities are predominant for $z/H < 0.2$. This behavior is associated with a current toward the bed (Figure 9c). At P1, vertical velocities are as the experiment E0 (Figure 9a). Thus, basically the image in time is as shown in Figure 7 in an instant, that is, the ejection is a quasi-steady feature of the flow.

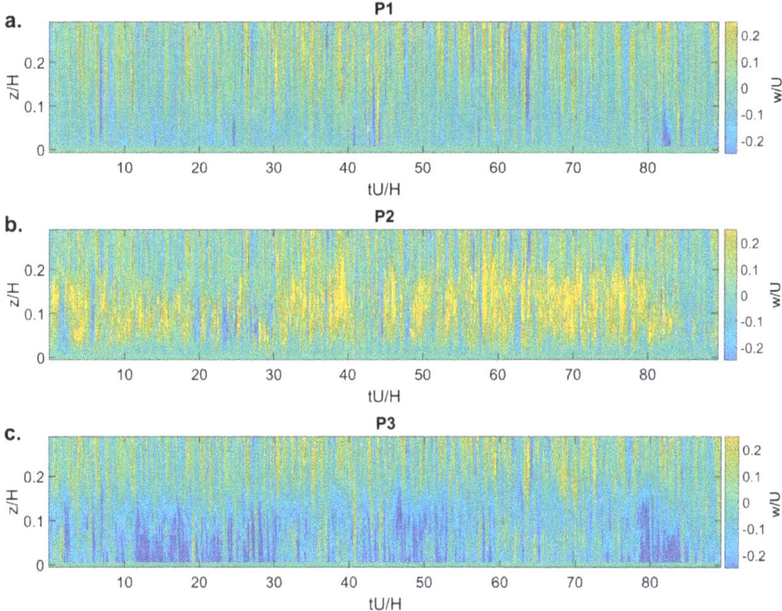

Figure 9. Vertical velocity made dimensionless with U, w/U in the experiment E1 after the spill of pumicite mixture for: (**a**) upstream of the center of the ejection, P1; (**b**) center of the ejection, P2 and (**c**) downstream from the center of the ejection, P3.

3.3. Scatter Plot of Velocity Fluctuations

The turbulence interactions associated with fine sediment ejections (E1) were compared with turbulence interaction without fine sediment (E0). In addition, the experiments measured with PIV in this article (P1, P2 and P3) were compared with Acoustic Doppler Velocimetry (ADV) measurements of turbulent interactions for an open channel with a gravel bed [38]. Experimental setup for the acquisition of ADV data of [38] was a flow rate of 14 L/s, a gravel bed of 45 mm particle diameter, a flow depth H of 0.1 m and $hm/H = 0.85$, where hm was the location of the ADV with respect to the bed. The distribution of the turbulence fluctuations (u', w') of an open channel in a Cartesian plane can be represented with an ellipse of negative slope and major axis (Ra) in the direction of quadrants 2 and 4 ($Q2 - Q4$) and minor axis (Rb) in direction of quadrants 1 and 3 ($Q1 - Q3$), as shown in Figure 10a,b, where the quadrants 1 to 4 are against clockwise circuit of the Cartesian plane and $Q1$ was ($u' > 0, w' < 0$). According to [12,16,17,19], the main turbulent coherent structures are ejections and sweeps, Q2 and Q4, respectively. The turbulent interactions in the narrow flume of this article have the same distribution as the widest flume of [38], i.e., an ellipse with a negative slope and the main turbulent coherent structures are ejections and sweeps; however, the magnitude of fluctuations are greater for [38], due to the turbulence is also greater, as shown Figure 10a,b. However, the pattern of those distributions change with respect to the turbulence in an open channel due to the presence of sediment ejections. These sediment ejections tend to increase the vertical fluctuations and decrease the streamwise fluctuations. These changes generate that outwards and inwards interactions can become events with a greater probability of occurrence than in a regular open channel, i.e., the ellipse has a positive slope and a relationship between *Ra* and *Rb* close to 1 (Figure 7c–e).

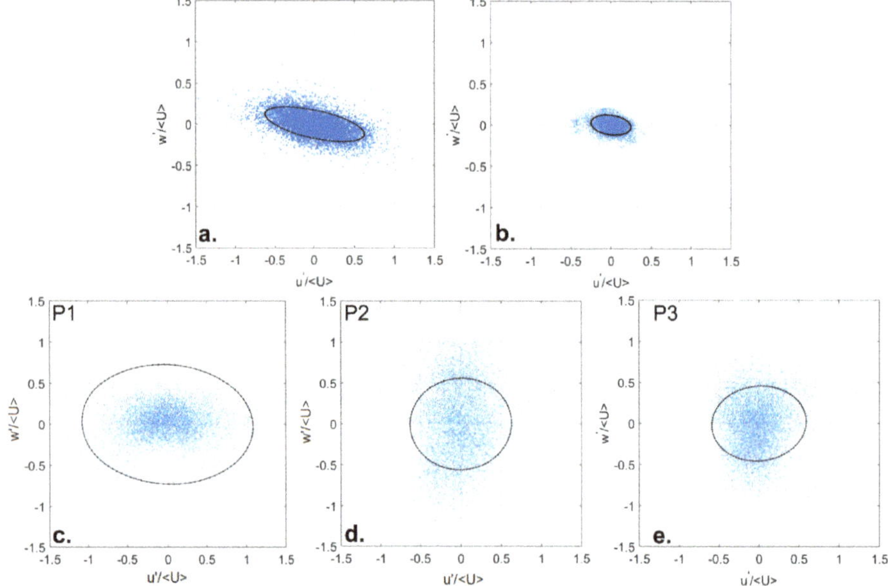

Figure 10. Scatter plot for: (**a**) ADV data [38], (**b**) PIV data in experiment E0, (**c**) PIV data in experiment E1 after the spill of pumicite mixture for P1, (**d**) PIV data in experiment E1 after the spill of pumicite mixture for P2 and (**e**) PIV data in experiment E1 after the spill of pumicite mixture for P3.

3.4. Frequency Spectra and Wavelet Analysis

Power spectrum frequency for vertical and streamwise velocities fluctuation are calculated both for E0 and E1 in points P1, P2, and P3 (Figure 11). Power spectrum was made dimensionless with $u_*^2 H$. In both cases, it is possible to identify the production zone and the inertial sub-range in the power spectrum for both cases before and after the spill of pumicite mixture. The inertial sub-range was considered between frequencies 10 and 20–30 Hz, where the slope of $-5/3$ is representative. The frequency of more than 20–30 Hz is noise in the PIV velocities. In the case of E0, in the production zone and in the three points of measurement, the energy is higher for streamwise velocity fluctuations than vertical velocity fluctuations. Conversely, for E1 after the spill of pumicite mixture, the vertical velocity fluctuation has higher energy in the production zone than the streamwise velocity fluctuation because the sediment ejection in P2 has dominant vertical velocities (Figure 11e). Then, according to the dynamics of sediment ejection (E1), where the vertical fluctuation is dominant, it can be seen that, at P2, the energy in the production zone for this components is highest, at P1 the energy in the production zone is the same as the experiment E0, and at P3 the energy in the production zone for the vertical component is large with respect to the streamwise component and can be considered as a downwelling point.

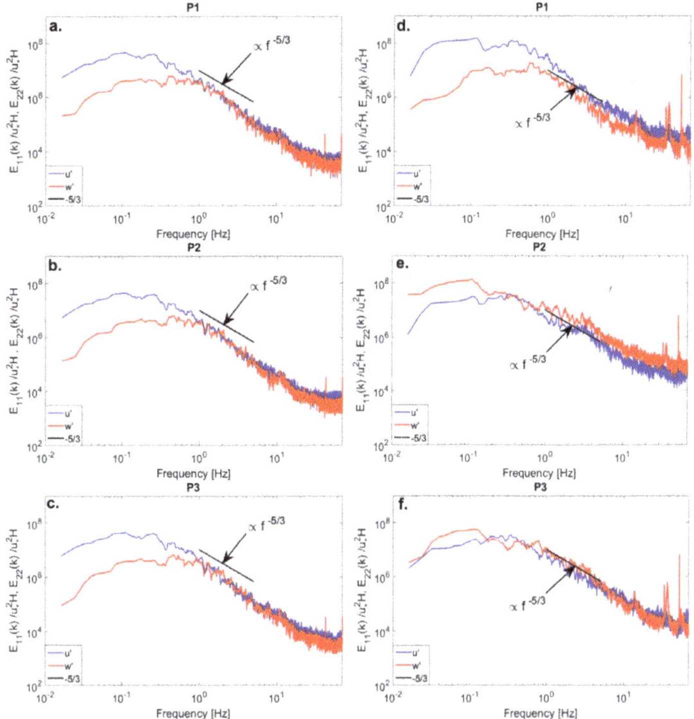

Figure 11. Power Spectrum Density for streamwise velocity fluctuation and vertical velocity fluctuations ($z/H = 0.1$), (**a–c**) in experiment E0, (**d–f**) in experiment E1, after the spill of fine sediment mixture.

Since the sediment ejection is a quasi-steady flow, then the local wavelet spectrum in the experiment E1, after the spill of fine sediment mixture, was implemented to analyze the u' and w' velocity components in the three points of the sediment ejection, P1, P2 and P3 (Figures 12 and 13). The wavelet spectra $|Wu|^2$ and $|Ww|^2$, corresponding to the wavelet spectra for u' and w' velocity fluctuations, respectively, were made dimensionless with dimensionless with Hu_*, as shown in Figures 12 and 13. The spectra have λ/H in vertical axes and tU/H in horizontal axes, where λ is the wavelength calculated as $\lambda = U/f$, f is frequency and tU/H is the dimensionless time. Wavelet analysis allows for seeing the variations of the power spectrum over time, i.e., with this analysis, a turbulent structure is determined and the time the structure is present in the measurement time. In the streamwise direction, upstream of sediment ejection, P1, the energy is concentrated for $\lambda = 10H$, for a long period of time, $10 < tU/H < 82$ (Figure 12). This wavelength is associated with a frequency of 0.06 Hz. According to the power spectrum shown in Figure 11e, it is a structure corresponding to a large-scale motion, i.e., this large-scale is present in almost all the time of the measurement time. In points P2 and P3, there is concentration of energy at this wavelength (frequency), but it is less intense than at P1, whereas, for $\lambda = 5H$ and $\lambda = 4H$, there is a concentration of energy at all the measurement points, P1 to P3. This wavelength is associated with frequencies of 0.13 Hz and 0.16 Hz, corresponding to large-scale motions (Figure 11d–f). However, this structure is present only at certain points in time; the period most frequent is $tU/H \sim 50$ (Figure 12a–c).

The wavelet spectrum for vertical fluctuation shows energy concentrations in P2 and P3 higher than those of P1; the highest concentration of energy is that of P2. This wavelet spectrum shows that, at point P2, during the entire measurement time, the vertical component of sediment ejection dominates, as shown in the spectrum of Figure 9b. According to the wavelet spectrum, Figure 13b,

$\lambda = 10H$ has a high energy concentration for $10 < tU/H < 82$. That wavelength corresponds to the frequency 0.06 Hz and is associated with a large scale of motion for P2 and P3 (Figure 11b,c). In addition, for $\lambda = 5H$, $4H$, and $3H$, there is a concentration of energy at points P2 and P3. This wavelength is associated with frequencies of 0.13 Hz, 0.16 Hz, and 0.21 Hz, respectively, corresponding to large-scale motions (Figure 11d–f). However, these structures are present only at certain points in time, the period most frequent is $tU/H \sim 40$ (Figure 12a–c).

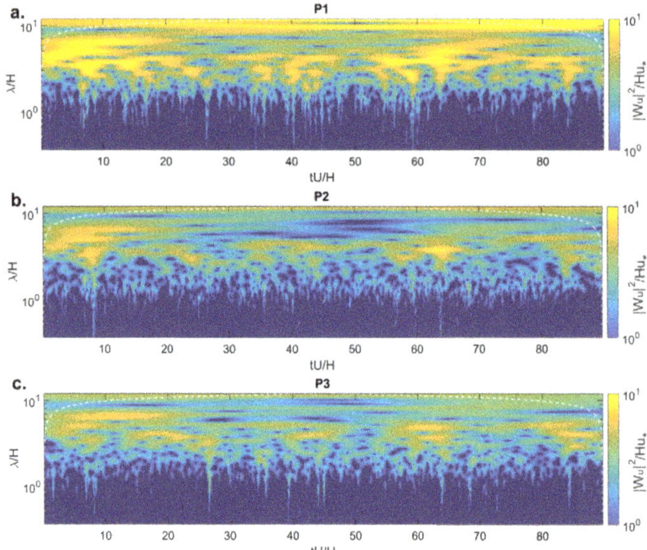

Figure 12. Wavelet spectrum for streamwise velocity fluctuations, u', in the experiment E1, after the spill of fine sediment mixture (**a**) P1, (**b**) P2, and (**c**) P3.

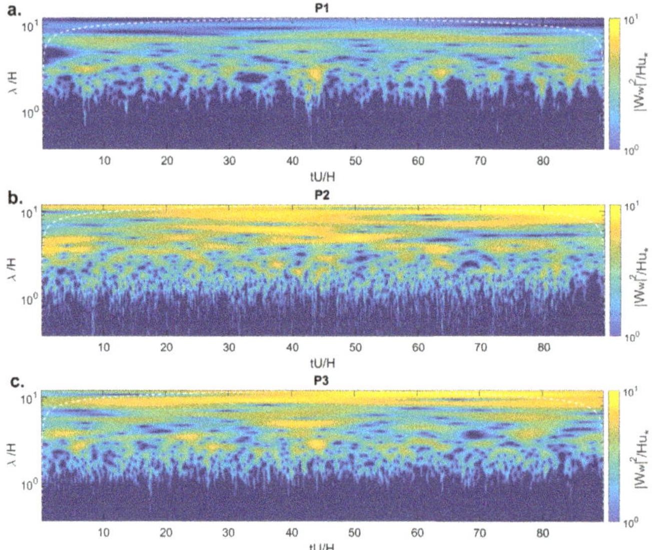

Figure 13. Wavelet spectrum for vertical velocity fluctuations, w', in the experiment E1, after the spill of fine sediment mixture, (**a**) P1, (**b**) P2, and (**c**) P3.

4. Discussion

The coherent structure in turbulence flows have been researched by [16,18,19,36,39] in open channels with permeable and impermeable beds, considering a steady flow. One methodology has been to analyze the Reynolds shear stress in quadrants of the Cartesian plane, with u' and w' and their distribution is described as an ellipse of negative slope and a turbulent event has a high probability to be in the quadrant $Q2$ and $Q4$, while $Q1$ and $Q3$ have a low probability of occurrence. In this study, we can see the same distribution of Reynolds shear stress in the case of open channel without fine sediment deposited into the bed (E0). However, the interaction between surface and subsurface flow in E1, after the spill of pumicite mixture, causes the analysis to change. The sediment ejection generates a quasi-steady flow from the bed toward the water column, where the vertical velocity component is higher than the streamwise velocity component, i.e., the turbulent interactions in $Q1$ and $Q3$ for that structure has a higher probability than in other research.

Furthermore, for isotropic turbulence in smooth open channel flow, Taylor's frozen turbulence hypothesis has been validated by [13,20,40] and others. Taylor's frozen turbulence hypothesis is also validated in experiment, E0, for $z/H > 0.1$. However, Ref. [41] recognizes the limitations of applying the Taylor's hypothesis. They considered an uncertainty for $z/H < 0.1$ because, in this region, the mean velocity and local velocity can diverge. Then, in experiment E0, the velocity difference between P1 and P2 for $z/H < 0.1$ is associated with a divergence of gravel pore velocities, so that, near the permeable bed, the Taylor's frozen turbulence hypothesis is invalidated, and these results are according to [41].

In experiment E1, fine sediment ejections are quasi-steady flows from the bed. Turbulence patterns change and turbulence becomes anisotropic because sediment ejection has a dominant vertical velocity. In this case, Taylor's frozen turbulence hypothesis is also invalidated.

The power spectrum density in open channels has been reported by [42–44]. They have validated the $-5/3$ slope in the measured spectra and show a greater spectral density of the streamwise component than the spanwise and vertical components. However, in this study, for the experiment E0, in the production zone, there is a greater spectral density of the streamwise component than the vertical component, but in the inertial subrange both components have a spectral density of similar magnitude. Conversely, in experiment E1, since the sediment ejection has a dominant upward movement, the spectral density changes. Upstream of the sediment ejection, at P1, in the production zone, the spectral density for streamwise component is greater than the vertical fluctuation component and the spectrum densities in both components are greater than the spectral density in experiment E0, whereas, in the center of the ejection, at P2, in the production zone and in the inertial subrange, the spectral density for vertical fluctuation component is greater than the spectral density for streamwise components. Downstream of the ejection in experiment E1, P3 can be considered as a downwelling point. It is important to note that sediment ejection increases the spectral energy, both in the production zone and in the inertial subrange.

Coherent structures of turbulence in open channels have been characterized by their sizes, such as small-scale motion, large-scale motion, and very-large scale motion by [13,20]. Streamwise scales of $3H$ and $10H$ allow coherent structures in these experiments to be classified as large-scale, i.e., hairpin vortex, and very large-scale motion, i.e., super streamwise vortex. Ref. [20] implemented the wavelet analysis to detect high concentrations of energy over time and their scale of motion, finding high concentrations of energy in $\lambda = 3H$ and $\lambda = 10H$, i.e., they reported hairpin vortices and super stream vortices; however, they did not show the influence cone, so the presence of the super stream vortices is not clear. The large-scale motion and the very large-scale motions find in this research are not the same coherent structures reported by [13,20] because the pattern of the sediment ejection has a dominant vertical component. In addition, the energy concentration changes with the measurement position inside the sediment ejection.

The interaction between ejections, sweeps, outward, and inward interactions generates turbulent structures such as horseshoe vortices, hairpin vortices, shedding vortices, etc. Those turbulent structures can move sediment away from where they occurred, i.e., they can generate erosion or scour [45]. Interactions such as ejections and sweeps are more common in open channels; however, in the experiments presented in this article, this is not entirely true. Outward and inward interactions are more frequent than in open channels. Furthermore, pumicite is a fine cohesive material, so mechanisms such as particle rolling, sliding, and saltation were not observed in the present experiments. Therefore, no erosion or scour was observed due to the sediment expulsion. However, there is a constant interaction between the surface and subsurface flow. This dynamic is relevant considering that the fine material can be, for example, mining materials and, therefore, a constant exchange of toxic material can be generated from the hyporheic zone to the surface flow.

The experimental scale is small compared with natural environments, but the dimensionless parameters in the experiments presented in this article on the basis of the flow and sediment transport phenomena in mountain streams are believed to be correct. The dimensionless parameter $H/D_g = 6.7$ of the experiments is representative of macroroughness flow and there is no bedload transport, that is, $\tau_* = 0.00006 < \tau_{*c} = 0.035$, where τ_* is the dimensionless shear stress, $\tau_* = u_*^2/(gRD_g)$, and τ_{*c} is the critical shear stress for the incipient motion of the sediment, with $R = (\rho_s - \rho)/\rho$, ρ_s and ρ are the sediment and water density, respectively [38]. However, the dimensionless parameter that is essential for this article is u_e/w_s, where u_e is the entrainment velocity and w_s is the fall velocity of the fine sediment [10]. The dimensionless parameter $u_e/w_s > 1$ would make entrainment possible. The flow average vertical velocity $< w >$ in P2 is 0.007 m/s (Figure 9) and w_s of the pumicite is 0.005 m/s, so, if u_e is equal to $< w >$, $< w > /w_s > 1$ and there is entrainment of the pumicite to the surface flow. The value of w/U in P2 observed in Figure 9 is in the order of 0.15, which shows that a significant part of the subsurface flow appears in the surface flow.

5. Conclusions

Fine sediment, such as pumicite, between the pores of a gravel river bed, can reduce the permeability and the initial porosity of the bed, modifying the roughness and hydraulic parameters of the subsurface flow. Pumicite has a low density, generating changes in the interaction between the subsurface and surface flow. These interactions are mass, momentum, and energy exchange, so the decrease of permeability of the gravel layer can generate an increase of vertical velocity and the turbulence intensities in the surface layer. Additionally, low velocities in the streamwise direction and high vertical velocities can break the streamwise structures associated with secondary currents near the bed and sediment ejections can be seen.

The sediment ejections change the patterns of turbulent structures and the distribution of the turbulence interactions, which means that the flow does not have a typical rough wall open channel flow turbulence. Additionally, the sediment ejections increase the energy both in the production zone and inertial subrange. Within the ejection, the vertical velocity component has the highest increase of energy in the center of the ejection. The sediment ejections could vary with the granular size of the subsurface layer and the density of the fine material, i.e., the low-density fine material in the subsurface layer encourages the presence of sediment ejections from the bed.

As future work, we will continue to evaluate the turbulent structures associated with sediment ejections in the presence of surface and subsurface flows. Fine sediments, with higher densities will be considered, for example, mining materials, such as tailings or metal concentrates, to evaluate the effect of particle density on the dynamics of the ejection. To characterize both spanwise and the sediment ejection in 3D, we will implement the technique Stereo PIV.

Author Contributions: Conceptualization, N.Y. and B.-P.N.; methodology, B.-P.N. and N.Y.; formal analysis, N.Y. and B.-P.N.; investigation, B.-P.N. and N.Y.; resources, N.Y.; writing—original draft preparation, B.-P.N.; writing—review and editing, N.Y.; supervision, N.Y.; project administration, N.Y.; funding acquisition, N.Y. In general, both authors contributed to the general development of the document through periodic meetings and research sessions. All authors have read and agreed to the published version of the manuscript.

Funding: The authors of this paper thank the financing of the Department of Civil Engineering, University of Chile, the Fondecyt Project 1140767, support from CONICYT through Beca Doctorado Nacional No 21181620 and Advanced Mining Technologic Center (AMTC) and CONICYT Project AFB180004.

Acknowledgments: The authors acknowledge Felipe Galaz for collaborating in conducting experiments.

Conflicts of Interest: The authors declare no conflict of interest.

Abbreviations

The following abbreviations are used in this manuscript:

Symbol table	Meaning
B	Wide flume.
D_g	Diameter of gravel.
D_s	Diameter of fine gravel.
D_p	Diameter of pumicite.
$E11$	Fourier Power spectrum streamwise velocity fluctuation.
$E22$	Fourier Power spectrum vertical velocity fluctuation.
f	Frequency.
Fr	Froude number.
H	Water depth.
K	Parameter of contribution to total Reynolds stress.
K'	Modified parameter of contribution to total Reynolds stress.
K'_{10}, K'_{90}	Values of K' associated with 10th and 90th percentiles, respectively.
Q_{sur}	Surface flow.
Q_{sub}	Subsurface flow.
R	$(\rho_s - \rho)/\rho$
Re	Reynolds number.
Ra, Rb	Major axis and minor axis of the ellipse.
S_f	Friction slope.
U	Section average mean velocity.
u	Instantaneous streamwise velocity.
w	Instantaneous vertical velocity.
$\bar{u}$,	Streamwise mean velocities.
$\bar{w}$	Vertical mean velocity.
u', w'	streamwise fluctuation, vertical fluctuation, respectively.
$<\bar{u}>$	Spatial average mean velocity for streamwise component.
$<\bar{w}>$	Spatial average mean velocity for vertical component.
U_{E0}	Depth average mean velocity in E0.
U_{E1}	Depth average mean velocity in E1.
$\tilde{u}$	Spacial variations of the time-averaged flow with respect to the double-averaged for streamwise component.
$\tilde{w}$	Spacial variations of the time-averaged flow with respect to the double-averaged for vertical component.
u_e	Entrainment velocity.
u_{*E0}	Shear velocity in E0.
u_{*E1}	Shear velocity in E1.
w_s	Fall velocity.
Wu	Wavelet coefficient streamwise velocity fluectuation.
Ww	Wavelet coefficient vertical velocity fluectuation.
α	Rotation angle of the ellipse.
λ	Wavelength.
ρ	Water density.
μ	kinematic viscosity.
τ_{tot}	Total shear stress.
τ_*	Dimensionless shear stress.
τ_{*c}	Critical shear stress.

References

1. Julien, P.Y. *Erosion and Sedimentation*, 2nd ed.; Cambridge University Press: New York, NY, USA, 2010; p. 371.
2. Cui, Y.; Parker, G. The arrested gravel front: Stable gravel-sand transitions in rivers Part 2: General numerical solution. *J. Hydraul. Res.* **1998**, *36*, 159–182. [CrossRef]
3. Beschta, R.; Jackson, W. The Intrusion of Fine Sediments into a Stable Gravel Bed. *J. Fish. Res. Board Can.* **1979**, *36*, 204–210. [CrossRef]
4. Diplas, P.; Parker, G. *Pollution of Gravel Spawning Grounds due to Fine Sediment*; Technical Report 240; Unversity of Minnesota: Minneapolis, MN, USA, 1985. [CrossRef]
5. Lisle, E. Correction to "Sediment Transport and Resulting Deposition in Spawning Gravels, North Coastal California" by Thomas E . Lisle. *Water Resour. Res.* **1989**, *25*, 1303–1319. [CrossRef]
6. Tonina, D.; Buffington, J.M. Hyporheic Exchange in Mountain Rivers I: Mechanics and Environmental Effects. *Geogr. Compass* **2009**, *3*, 1063–1086. [CrossRef]
7. Bagnold, R.A. An Approach to the Sediment Transport Problem from General Physics. *USGS Prof. Pap.* **1966**, *42*. [CrossRef]
8. Jackson, R.G. Sedimentological and fluid-dynamic implications of the turbulent bursting phenomenon in geophysical flows. *J. Fluid Mech.* **1976**, *77*, 531–560. [CrossRef]
9. Niño, Y.; Garcia, M.H. Experiments on particle—Turbulence interactions in the near—Wall region of an open channel flow: Implications for sediment transport. *J. Fluid Mech.* **1996**, *326*, 285–319. [CrossRef]
10. Niño, Y.; Lopez, F.; Garcia, M. Threshold for particle entrainment into suspension. *Sedimentology* **2003**, *50*, 247–263. [CrossRef]
11. Tamburrino, A.; Gulliver, J.S. Large flow structures in a turbulent open channel flow. *J. Hydraul. Res.* **1999**, *37*, 363–380. [CrossRef]
12. Adrian, R.J. Hairpin vortex organization in wall turbulence. *Phys. Fluids* **2007**, *19*. [CrossRef]
13. Zhong, Q.; Li, D.; Chen, Q.; Wang, X. Coherent structures and their interactions in smooth open channel flows. *Environ. Fluid Mech.* **2015**, *15*, 653–672. [CrossRef]
14. Zhong, Q.; Chen, Q.; Wang, H.; Li, D.; Wang, X. Statistical analysis of turbulent super-streamwise vortices based on observations of streaky structures near the free surface in the smooth open channel flow. *J. Am. Water Resour. Assoc.* **2016**, *5*, 2. [CrossRef]
15. Nakagawa, H.; Tsujimoto, T.; Shimizu, Y. Turbulent flow with small relative submergence. In *Fluvial Hydraulics of Mountain Regions*; Springer: Berlin/Heidelberg, Germany, 1991; pp. 33–44.
16. Niño, Y.; Musalem, R. Turbulent entrainment events of sediment grains over bedforms. In Proceedings of the Fourteenth Engineering Mechanics Conference, Austin, TX, USA, 21–24 May 2000; The University of Texas at Austin: Austin, TX, USA, 2000.
17. Sambrook Smith, G.H.; Nicholas, A.P. Effect on flow structure of sand deposition on a gravel bed: Results from a two-dimensional flume experiment. *Water Resour. Res.* **2005**, *41*, 1–12. [CrossRef]
18. Cooper, J.R.; Ockleford, A.; Rice, S.P.; Powell, D.M. Does the permeability of gravel river beds affect near-bed hydrodynamics? *Earth Surf. Process. Landf.* **2018**, *43*, 943–955. [CrossRef]
19. Wallace, J.M. Quadrant Analysis in Turbulence Research: History and Evolution. *Annu. Rev. Fluid Mech.* **2016**, *48*, 131–158. [CrossRef]
20. Chen, K.; Zhang, Y.; Zhong, Q. Wavelet coherency structure in open channel flow. *Water* **2019**, *11*, 1664. [CrossRef]
21. Roussinova, V. Turbulent Structures in Smooth and Rough Open Channel Flows: Effect of Depth. Ph.D. Thesis, University of Windsor, Windsor, ON, Canada, 2009.
22. Manes, C.; Pokrajac, D.; McEwan, I.; Nikora, V. Turbulence structure of open channel flows over permeable and impermeable beds: A comparative study. *Phys. Fluids* **2009**, *21*, 1–12. [CrossRef]
23. Bailly, C.; Comte-Bellot, G. Turbulence. In *Experiments Fluid Mechanics*; Springer: Berlin, Germany, 2015; p. 360.
24. García, C.M.; García, M.H. Characterization of flow turbulence in large-scale bubble-plume experiments. *Exp. Fluids* **2006**, *41*, 91–101. [CrossRef]
25. Bronshtein, I.N.; Semendyayev, K.; Musiol, G.; Muehlig, H. *Handbook of Mathematics*, 5th ed.; Springer: New York, NY, USA, 2007; pp. 741–743.

26. Addison, P.S.; Watson, J.N.; Feng, T. Low-oscillation complex wavelets. *J. Sound Vib.* **2002**, *254*, 733–762. [CrossRef]

27. Bendat, J.S.; Piersol, A.G. *Random Data: Analysis and Measurement*, 4th ed.; Wiley: New York, NY, USA, 2010; p. 604.

28. Grinsted, A.; Moore, J.C.; Jevrejeva, S. Application of the cross wavelet transform and wavelet coherence to geophysical time series. *Nonlinear Process. Geophys.* **2004**, *11*, 561–566. [CrossRef]

29. Bustamante-Penagos, N.; Niño, Y. Percolation of the Copper Concentrate in a Porous Medium. In Proceedings of the XXVIII Congreso Latinoamericano de Hidráulica, Buenos Aires, Argentina, 18–21 September 2018; Lecertúa, E., Lopardo, M.C., Menéndez, Á., Spalletti, P., Eds.; IAHR: Panama City, Panama, **2018**; pp. 2076–2084. (In Spanish)

30. Bustamante-Penagos, N.; Niño, Y. Copper concentrate and gravel beds. In Proceedings of the 38th IAHR World Congress, Panama City, Panama, 1–6 September 2019; Calvo, L., Ed.; IAHR: Panama City, Panama, 2019; pp. 5080–5090. [CrossRef]

31. Nikora, V.; Koll, K.; McEwan, I.; McLean, S.; Dittrich, A. Velocity distribution in the roughness layer of rough-bed flows. *J. Hydraul. Eng.* **2004**, *130*, 1036–1042. [CrossRef]

32. Szlapczynski, R.; Szlapczynska, J. An analysis of domain-based ship collision risk parameters. *Ocean Eng.* **2016**, *126*, 47–56. [CrossRef]

33. Torrence, C.; Compo, G. A practical guide to wavelet analysis. *Bull. Am. Meteorol. Soc.* **1998**, *79*, 61–78. [CrossRef]

34. Cohen, E.A.; Walden, A.T. A statistical analysis of Morse wavelet coherence. *IEEE Trans. Signal Process.* **2010**, *58*, 980–989. [CrossRef]

35. Nikora, V.; Goring, D.; McEwan, I.; Griffiths, G. Spatially Averaged Open-Channel Flow Over Rough Bed. *J. Hydraul. Eng.* **2001**, *127*, 123–133. [CrossRef]

36. Hofland, B. Rock & Roll: Turbulence-Induced Damage to Granular Bed Protections. Ph.D. Thesis, Delft University of Technology, Delft, The Netherlands, 2005.

37. Wren, D.G.; Langendoen, E.J.; Kuhnle, R.A. Effects of sand addition on turbulent flow over an immobile gravel bed. *J. Geophys. Res. Earth Surf.* **2011**, *116*, 1–12. [CrossRef]

38. Niño, Y.; Licanqueo, W.; Janampa, C.; Tamburrino, A. Front of unimpeded infiltrated sand moving as sediment transport through immobile coarse gravel. *J. Hydraul. Res.* **2018**, *1686*, 18. [CrossRef]

39. Lichtner, D.T. Turbulent Interactions between Stream Flow and Near-Subsurface Flow: A Laboratory Approach Using Particle Image Velocimetry and Refractive Index Matching. Master's Thesis, University of Illinois at Urbana-Champaign, Champaign, IL, USA, 2015.

40. Cenedese, A.; Romano, G.P.; Di Felice, F. Experimental testing of Taylor's hypothesis by L.D.A. in highly turbulent flow. *Exp. Fluids* **1991**, *11*, 351–358. [CrossRef]

41. Cameron, S.M.; Nikora, V.I.; Stewart, M.T. Very-large-scale motions in rough-bed open-channel flow. *J. Fluid Mech.* **2017**, *814*, 416–429. [CrossRef]

42. García, C.M.; Cantero, M.I.; Niño, Y.; García, M.H. Turbulence Measurements with Acoustic Doppler Velocimeters. *J. Hydraul. Eng.* **2005**, *131*, 1062–1073. [CrossRef]

43. Nikora, V.; Nokes, R.; Veale, W.; Davidson, M.; Jirka, G.H. Large-scale turbulent structure of uniform shallow free-surface flows. *Environ. Fluid Mech.* **2007**, *7*, 159–172. [CrossRef]

44. Balakumar, B.J.; Adrian, R.J. Large- and very-large-scale motions in channel and boundary-layer flows. *Philos. Trans. R. Soc. A Math. Phys. Eng. Sci.* **2007**, *365*, 665–681. [CrossRef] [PubMed]

45. Dey, S. *Fluvial Hydrodynamics*; Springer: Berlin, Germany, 2014; p. 706. [CrossRef]

Article

Wavelet Coherency Structure in Open Channel Flow

Kebing Chen [1], Yifan Zhang [1] and Qiang Zhong [1,2,*]

[1] College of Water Resources and Civil Engineering, China Agricultural University, Beijing 100083, China
[2] Beijing Engineering Research Center of Safety and Energy Saving Technology for Water Supply Network System, China Agricultural University, Beijing 100083, China
* Correspondence: qzhong@cau.edu.cn

Received: 4 July 2019; Accepted: 6 August 2019; Published: 12 August 2019

Abstract: Many studies based on single-point measurement data have demonstrated the impressive ability of wavelet coherency analysis to catch the coherent structures in the wall-bounded flows; however, the question of how the events found by the wavelet coherency analysis relate to the features of the coherent structures remains open. Time series of velocity fields in x–y plane of the steady open channel flow was obtained from a time-resolved particle imaging velocimetry system. The local wavelet spectrum found shows that one of the main energetic scales in open channel flows is $3h$ motions. The wavelet coherent coefficients of u and v series from the same measurement points successfully detected the occurrence and the scale of these $3h$ motions, and the phase angle indicates their inside velocity structure is organized by the Q2 and Q4 events. The wavelet coherency analysis between different measurement points further reveals the incline feature of the $3h$ scale motions. All the features of this $3h$ motion found by the wavelet coherency analysis coincide well with the flow field that is created by the passing of hairpin packets.

Keywords: wavelet coherency; Taylor's frozen turbulence hypothesis; scale; hairpin vortex packet; open channel flow

1. Introduction

The essential difficulties of turbulence are the numerous scale motions and the complex interactions between them. Therefore, the two major tasks of turbulent data analysis can be summarized: The first one is scale decomposition, which decomposes the different scale components from the original turbulent signal; the other one is the correlation of scales, which analyzes the interactions between different scales.

Traditional turbulent data analysis is based on the Fourier transform. Fourier analysis decomposes the turbulent signal into the sum of infinite sine and cosine functions with different frequencies. It can reveal the frequency characteristic of the turbulent signal, but it cannot determine when those frequency components occurred. This is due to the fundamental difference between the presuppositions of Fourier analysis and intermittency of the turbulent signal. Specifically, each frequency in Fourier analysis stands for one scale vortices, and the passing of one vortex will cause one period sine wave at the measurement point. The continuous sine waves from positive to negative infinity on the time axis make Fourier analysis implicitly presuppose that each scale vortices occupy the entire timeline. However, the vortices in turbulence occur intermittently, and they present in the portion of both time and space axis and only effect the local flow field.

Wavelet analysis overcomes this defect by using the waves fluctuating in the limited time as the basis functions, and the behavior of the signal at infinity does not play any role. Thus, wavelet analysis is performed in both time and frequency domain, allowing the definition of local spectral properties and the ability to zoom in on local features of the signal. Since Farge [1] introduced the wavelet analysis into turbulence, wavelets have become pervasive in turbulent signal analysis. Most of the applications use the wavelet analysis only as a scale decomposition tool, and the research on the correlation of scales are still based on the traditional correlation coefficient or the Fourier coherency analysis.

Following the definition of coherency in Fourier analysis, Liu [2] firstly defined the wavelet coherency and applied the signal from ocean wind waves. The wavelet coherency analysis can reveal the similarity of two signals at different scales and instants, which provides a powerful tool to reveal the correlation between two signals of different types or from different locations which is cased by the local coherent structures. The fundamentals of the cross-wavelet and wavelet coherency analysis were systematically introduced by Torrence and Compo [3], and Torrence and Webster [4] proposed a broadly applicable smoothing function for the wavelet coherency, which made wavelet coherency eventually become a universal data analysis method. Grinsted et al. [5] developed a software package that allows users to perform the cross wavelet transform and wavelet coherency analysis, and they applied the methods to the Arctic Oscillation index and the Baltic maximum sea ice extent record. Camussi et al. [6] applied the cross wavelet transform and wavelet coherency analysis to wall pressure fluctuation signals from a microphone pair in the incompressible turbulent boundary layers. The time instants corresponding to a local wavelet coherency overcoming the trigger level were selected as the coherent events, and conditional statistical analysis was performed on these selected events to show the physical features of these highly coherent events. Based on the conditional statistics, the highly coherent event was speculated as the footprint of the near-wall-sweep-type motion which are known to be closely associated with the presence of streamwise vortices embedded within the turbulent flow and located in the near-wall region.

From the previous literature review, it can be seen that most of the applications of the wavelet coherency was done on single point measurement signals. These single point signals contain the information of the events that happen in turbulence, but they cannot reflect the whole picture of the events, and the question of how these highly coherent events revealed by the wavelet coherent analysis from single point measurement data related to the two or three dimensional coherent structures in the flow field remains open. Although we can easily get two- or even three-dimensional flow fields in the laboratory today, a large number of tasks in the actual application scenario are still completed by single-point measurement. Therefore, it is instructive for the single point data analysis to explore the three-dimensional coherent structures corresponding to events in them. Present study employs the time-resolved particle imaging velocimetry system (TR-PIV) to get the two dimensional flow field time series, which makes it possible to find out the flow structures from the 2D flow fields at the corresponding wavelet coherent events from the single point time series. The following part of this paper is organized as follows. Section 1 describes the principle of the wavelet coherency analysis briefly. Section 2 introduces the experiment and data. Section 3 presents the results of wavelet spectral and coherency analysis. Section 4 summarizes major findings.

2. Wavelet Coherency

We limit ourselves to a brief introduction of the wavelet coherency. The detailed procedures can be found in [3–5]. The continuous wavelet transform of a velocity signal $u(t)$ is defined as the convolution of $u(t)$ with a scaled and normalized wavelet:

$$W_u(a, t_0) = \int_{-\infty}^{+\infty} \left[u(t') \frac{1}{\sqrt{a}} \psi^* \left(\frac{t_0 - t'}{a} \right) \right] dt' \tag{1}$$

where ψ is the wavelet function; superscript * denotes the complex conjugate; a and t_0 are the scale and position parameter, respectively; $W_u(a, t_0)$ is the wavelet transform coefficient of u at scale a and instant t_0. The wavelet function ψ in the wavelet transform must have zero mean and be localized in both time and frequency domain [1]. Following Liu [2], Torrence and Compo [3] and Chui [7], the complex Morlet wavelet is used in this study since it provides a good balance between time and frequency localization:

$$\psi(t) = \pi^{-1/4} e^{i\omega_0 t} e^{-\frac{t^2}{2}} \tag{2}$$

where ω_0 is a wavelet center frequency, here taken to be 6 to balance between time and frequency localization. From Equation (1), it can be seen that the wavelet function is stretched and shifted in time by varying the scale and position parameter. Thus, different scale motions at every instant can be separate out from the original signal by the wavelet transform. In order to facilitate comparison, the pseudo Fourier frequency corresponding to the scale a is usually determined by picking up the energy peak in the Fourier spectrum of the wavelet function. For the Morlet wavelet with $\omega_0 = 6$, the Fourier frequency f is almost equal to the scale, e.g., Fourier frequency f_0 is 0.971Hz when $a = 1$, and the corresponding frequency f at scale a is determined by:

$$f = \frac{f_0}{a} \tag{3}$$

When the pseudo Fourier frequency f is determined, the wave number k and wavelength λ can be obtained based on Taylor's hypothesis:

$$\begin{cases} k = \frac{2\pi f}{U} \\ \lambda = \frac{2\pi}{k} \end{cases} \tag{4}$$

where U is the mean streamwise velocity at that point.

Similar to Fourier analysis, the cross wavelet coefficient can be obtained:

$$W_{uv}(a,t) = W_u(a,t) W_v^*(a,t) \tag{5}$$

There is a Parseval relation [8]:

$$\int_{-\infty}^{+\infty} [u(t)v^*(t)]dt = \frac{1}{c_\psi} \int_{-\infty}^{+\infty} \int_0^{+\infty} W_u(a,t) W_v^*(a,t) \frac{da}{a^2} dt = \frac{1}{c_\psi} \int_{-\infty}^{+\infty} \int_0^{+\infty} W_{uv}(a,t) \frac{da}{a^2} dt \tag{6}$$

where c_ψ is a factor depended by the wavelet function:

$$\begin{cases} c_\psi = \int_0^{\infty} |\hat{\psi}(\omega)|^2 \frac{d\omega}{|\omega|} \\ \hat{\psi}(\omega) = \int_{-\infty}^{+\infty} \psi(t) e^{-2i\pi\omega t} dt \end{cases} \tag{7}$$

In the turbulence situation, the velocity $u(t)$ and $v(t)$ are both real signal. Thus, we have:

$$\begin{aligned} E_{total} = \int_{-\infty}^{+\infty} u(t)^2 dt \ &= \frac{1}{c_\psi} \int_{-\infty}^{+\infty} \int_0^{+\infty} |W_{uu}(a,t)|^2 \frac{da}{a^2} dt \\ &= \frac{1}{c_\psi k_0} \int_{-\infty}^{+\infty} \int_0^{+\infty} \left|W_{uu}\left(\frac{f_0}{f},t\right)\right|^2 df dt \end{aligned} \tag{8}$$

where E_{total} can be considered as the turbulent kinetic energy. Equation (8) shows that the wavelet transform coefficients actually represent the energy carried by the corresponding scale motions at the given instant. Thus $|W_{uu}(f,t)|^2$ is called local wavelet spectrum. The frequency spectrum can be defined as:

$$E_{uu}(f) = \frac{1}{C} \int_{-\infty}^{+\infty} |W_{uu}(f,t)|^2 dt \tag{9}$$

where C is the normalization constant which makes

$$\int_0^{+\infty} E_{uu}(f)df = u'^2 \tag{10}$$

in which u' is the standard deviation of time series $u(t)$.

For the correlation coefficient, we have the definition:

$$R_{uv} = \frac{\int_{-\infty}^{+\infty}[u(t)v(t)]dt}{\sqrt{\int_{-\infty}^{+\infty}u(t)^2dt} \cdot \sqrt{\int_{-\infty}^{+\infty}v(t)^2dt}} \tag{11}$$

Based on Equation (6), we can get:

$$R_{uv}^2 = \frac{\mathfrak{R}\left[\int_{-\infty}^{+\infty}\int_0^{+\infty} W_{uv}(a,t)\frac{da}{a^2}dt\right]^2}{\left[\int_{-\infty}^{+\infty}\int_0^{+\infty}|W_u(a,t)|^2\frac{da}{a^2}dt\right]\cdot\left[\int_{-\infty}^{+\infty}\int_0^{+\infty}|W_v(a,t)|^2\frac{da}{a^2}dt\right]} \tag{12}$$

in which $\mathfrak{R}[\cdot]$ is the real part of complex variables. Similar to the Equation (12), we can define the wavelet coherency as

$$R_{uv}^2(a,t) = \frac{\mathfrak{R}[W_{uv}(a,t)]^2}{|W_u(a,t)|^2 \cdot |W_v(a,t)|^2} \tag{13}$$

However, the correlation between two signals at one instant obviously makes no sense. As noted by Liu [2], this coherency is identically one at all times and scales. This problem is circumvented by smoothing the wavelet coefficient along time or both time and scale axis before normalizing. The time and scale smoothing operator given by Torrence and Webster [4] is used in this study:

$$\begin{cases} R_{uv}^2(a,t) = \dfrac{\left|S[a^{-1}W_{uv}(a,t)]\right|^2}{S[a^{-1}|W_u(a,t)|^2]\cdot S[a^{-1}|W_v(a,t)|^2]} \\ S[W] = S_{scale}(S_{time}(W)) \\ S_{scale}(W(a,t)) = W(a,t) * c_1\Pi(0.6a) \\ S_{time}(W(a,t)) = W(a,t) * c_2 e^{\frac{-t^2}{2a^2}} \end{cases} \tag{14}$$

in which the symbol * denotes the convolution product; $\Pi(0.6a)$ is a boxcar filter of width 0.6; $e^{\frac{-t^2}{2a^2}}$ is the absolute value of the Morlet wavelet; c_1 and c_2 are normalization coefficients to have a total weight of unity. The factor of 0.6 is the empirically determined scale decorrelation length for the Morlet wavelet. The wavelet-coherence phase difference is given by:

$$\phi(a,t) = \arctan\left\{\frac{\mathfrak{I}[S[a^{-1}W_{uv}(a,t)]]}{\mathfrak{R}[S[a^{-1}W_{uv}(a,t)]]}\right\} \tag{15}$$

in which $\mathfrak{I}[\cdot]$ is the imaginary part of the complex variables.

The MATLAB function "wcoherence" is used to compute the wavelet coherency between two signals in this study. One can find the wavelet coherency of analytical signals in the documentation of the MATLAB function, which reveals the ability of wavelet coherency to analyze the scale and phase relations between two signals.

3. Experiment

Experiments were conducted in the Tsinghua tilting hydraulic flume, which is a closed-circuit open channel 20 m long and 0.3 m wide. The measurement section was set up 12 m downstream of the flume entrance. Eight ultrasonic water level sensors were set on the flume to monitor the water depth across the entire flume. The streamwise and wall-normal directions are denoted by x and y, respectively, and the corresponding components of fluctuating velocities are u and v. Velocity field measurements of the x–y plane in the middle of channel were made at a uniform flow condition as listed in Table 1. The water depth h is 2.9 cm. Thus, the flow in the central region can be considered as statistically two-dimensional [9,10]. The Reynolds number based on the section average velocity U_m and water depth was 15,895 (U_m was based on the discharge from the electromagnetic flowmeter), and the friction Reynolds number was 880 ($u_* = \sqrt{ghJ}$).

Table 1. Flow condition and particle imaging velocimetry (PIV) parameters.

J	h	ν	B/h	U_m	u^*	Fr	Re	Re_τ
-	(cm)	$(10^{-2}$ cm^2/s)	-		(cm/s)	-	-	-
0.0036	2.90	1.06	10.3	58.1	3.29	1.09	15,895	880

Image size pixels	Exposure time μs	Frequency Hz	Resolution pixels/mm	Number of images
1280 × 896	150	2500	32	5596

J = bed slope, h = water depth, ν = kinematic viscosity, B = channel width, B/h = aspect ratio, U_m = the depth-averaged velocity, u^* = friction velocity, Fr = Froude number, Re = Reynolds number, Re_τ = friction Reynolds number.

Instantaneous, two-dimensional velocity fields were measured in the streamwise-wall-normal plane (x–y plane) with a TR-PIV. The laser sheet was projected from the channel bed and it was located at the central line of the channel. The camera was set at the side of the channel and the laser sheet plane and the CMOS plane of the camera kept parallel to the mid-vertical plane of the channel. The PIV parameters are also listed in the Table 1. The exposure time was fixed at 150 μs as a compromise between minimizing image streaking and maximizing image lightness. The sampling frequency was 2500 Hz to obtain time-resolved series of 2D flow fields. A total number of 5596 images were obtained thus the time series contains 5595 continuous velocity fields. A greater sample number will be better but the capacity of the high-speed memory in the high-speed camera limits the number of images. Particle images were analyzed by using the iterative multi-grid image deformation method. Various test results of the PIV algorithm used in this paper can be found in the report of the 4th International PIV Challenge (the symbol of the PIV algorithm is TsU) [11]. The window size in the final iterative step is 16 × 16 pixels with a 50% overlap. A detailed description of the experiment system can be found in [12,13].

4. Results

4.1. Wavelet Spectrum

Figure 1a presents the wavenumber spectrum E_{uu} for streamwise velocity fluctuation at point ($x/h = 0$, $y/h = 0.5$) from both Fourier and wavelet analysis. The biggest wavelength was chosen as $10h$ because the length of the time series is $tU/h = 45.5$, and this length is not enough for obtaining a creditable result for the larger scale motions. It can be seen from Figure 1a that the wavelet spectrum is much smoother and follow the Fourier spectrum well. This can be attributed to the fact that the wavelet spectrum presents an average of the Fourier spectrum weighted by the square of the Fourier transform of the analyzing wavelet shifted at wave number k, and it keeps the same power-law as in the Fourier spectrum [14]. Owing to the relatively small sample size, even the wavelet spectrum is still spiky. The TR-PIV system can capture the 2D flow field time series, but it is very hard to get big sample sizes because of the storage and transport difficulties of the huge amount of data. However, the main goal of this study is to show the physical meaning revealed by the wavelet analysis on the turbulent signal from open channel flows instead of presenting accurate wavenumber spectra.

Figure 1b shows the pre-multiplied power spectrum, $kE_{uu}(k)$, at point ($x/h = 0$, $y/h = 0.5$) from the wavelet spectrum. Based on Equation (10), when the horizontal axis is set to logarithmic coordinates, we have

$$u'^2 = \int_0^{+\infty} E_{uu}(k)dk = \int_0^{+\infty} kE_{uu}(k)d(\ln k) \tag{16}$$

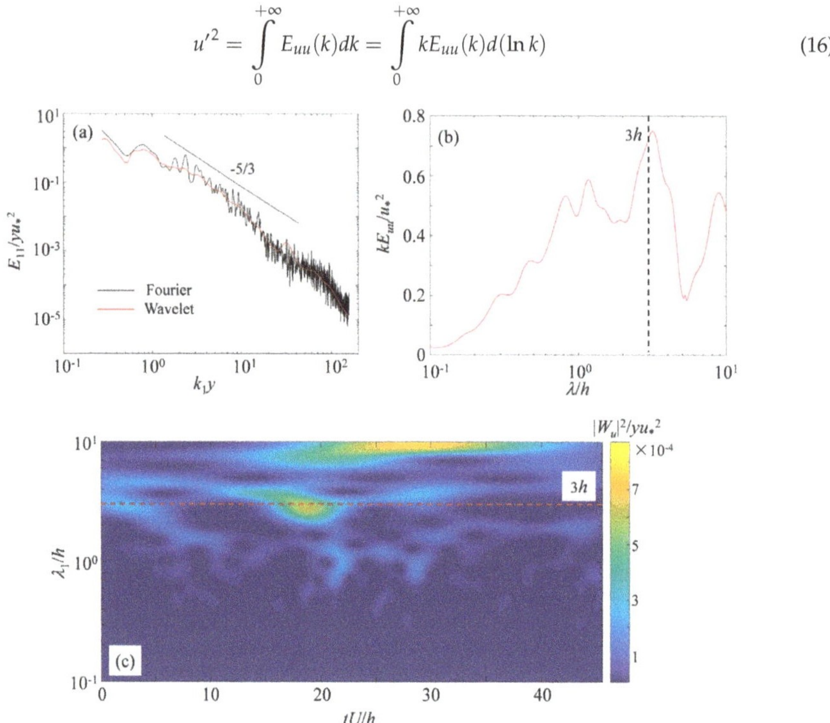

Figure 1. The wavenumber spectrum (**a**), pre-multiplied spectrum (**b**) and local wavelet spectrum (**c**) of the streamwise velocity series at the measurement point ($x/h = 0$, $y/h = 0.5$). The black solid straight line in (**a**) shows the classical −5/3 law in the turbulence. The black dash line in (**b**) and red dash line in (**c**) is marked the strongest scale in the spectrum.

Therefore, the whole area under the pre-multiplied spectrum curve in the semi-log plot is directly related to the value of turbulence intensity, and the area under a small section of the spectrum curve can be considered as the strength of the corresponding scale motions. The strongest scale in Figure 1b is approximately $3h$ as marked by the black dash line, which means the $3h$ scale motions are one of the main energetic structures in the outer layer. In the scales greater than $3h$, there may exist another energetic mode around $10h$. This two-energetic-mode feature of the pre-multiplied spectrum is similar to the results from other wall-bounded flows [15–17]. By the coherent structure classification in open channel flow [12,13], the h and $10h$ order motions can be classified as large- and very-large-scale structures.

The traditional spectrum gives the general information about the energetic scales but cannot show the time instants when the local event happens. The contour map of the local wavelet spectrum, $|W_u|^2$ is shown in Figure 1c. There are two high energy regions at scale approximately $3h$ and $10h$, respectively, which coincides with the result in pre-multiplied spectrum in Figure 1b. From $|W_u|^2$, we can see that the strongest $3h$ and $10h$ scale motions pass the measurement point during $17 < tU/h < 21$ and $25 < tU/h < 35$, respectively.

In order to show the entire flow structures when the local wavelet spectrum shows high energy, fluctuating velocity field series from $tU/h = 11$ to 35 are pieced together based on Taylor's frozen turbulence hypothesis in Figure 2, following Zhong et al. [12].

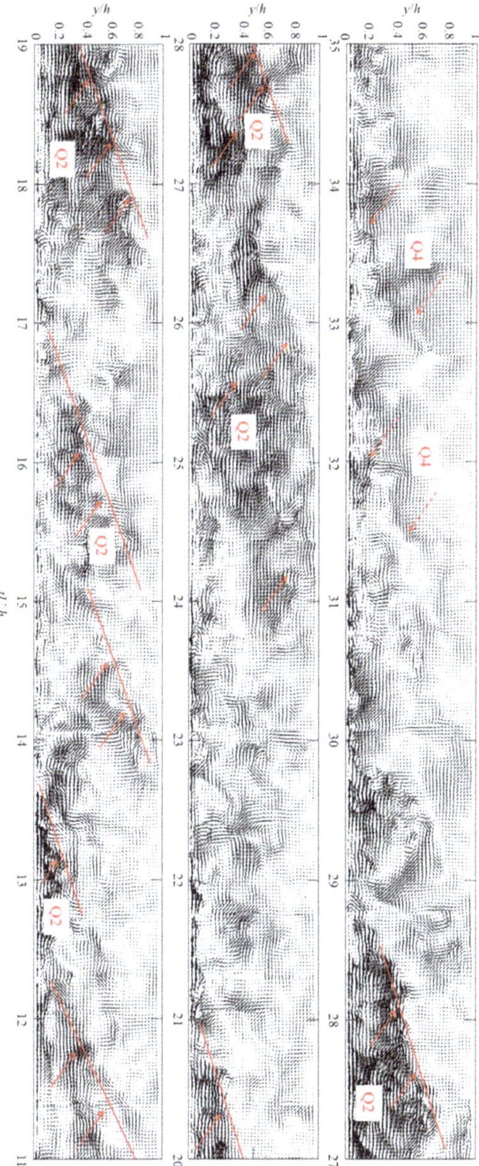

Figure 2. Fluctuating velocity fields pieced together based on Taylor's frozen turbulence hypothesis. The red arrows indicate the strong Q2 and Q4 events in the flow field, and the red solid lines indicate the inclined shear layers.

The convection velocity in Taylor's frozen turbulence hypothesis is the mean velocity of the whole field. The major characteristic of flow field during $17 < tU/h < 21$ is an inclined shear layer (marked by the red solid lines in Figure 2) with the strong ejections (marked by the red solid arrows in Figure 2). The ejections are also called the Q2 event, where the fluctuating streamwise velocity $u < 0$ and vertical velocity $v > 0$. When $u > 0$ and vertical velocity $v < 0$, there is a sweep motion, or Q4 event. The streamwise scale of this inclined shear layer structure is approximately $2.5h$. The features of this inclined shear layer structure are similar to the flow field in the reference [12,18–20], and it is usually considered as the typical sign of the passing of a hairpin packet [19,21]. The transport effect of the hairpin vortices in the packet induce the strong Q2 events, and the shear line is inclined because the heads of hairpin vortices in the packet describe an envelope inclined at 15–20° with respect to the wall. Considering the information from flow field and the results in previous literature, the highly energetic region at $3h$ scale in the wavelet spectrum reveals the passing of the hairpin packet. In Figure 2 it can be seen that from $tU/h = 11$ to 30 there are at least six typical inclined shear layers located almost end to end, and the wavelet spectrum indeed shows high energy in the corresponding duration and scale.

For the $10h$ scale highly energetic region, the flow field during $25 < tU/h < 35$ shows hairpin packets firstly and then strong Q4 events (marked by the red dash arrows in Figure 2). It indicates that the $10h$ order motions consist of two different types of smaller structures, hairpin packets, and Q4 events. This agrees with the phenomena reported by Zhong et al. [12], Adrian and Marusic [22], Zhong et al. [12] and Zhong et al. [13] suggested a super-streamwise vortex model for the $10h$ order motions in open channel flows. The super-streamwise vortices rotate around the x axis. The strong Q2 events from hairpin packets constitute the upward movement and the Q4 events are the downward movement of the super-streamwise vortices.

From the above discussion, the energetic scales can be presented by the traditional spectrum analysis, and the wavelet spectrum can show not only the energetic scales but also the moment these energetic structures occur. However, wavelet spectrum contains no information about the organization of these structures. The following analysis will show that the wavelet coherency can reveal more details about the energetic structures.

4.2. Wavelet Coherency

The wavelet coherency between time series u and v at the same measurement point ($x/h = 0$, $y/h = 0.5$) is shown in Figure 3. The advection velocity U is the mean velocity of the whole field. The white dash line marks the cone of influence of the boundaries. The biggest wavelength was chosen as $10h$ as in Figure 1. It can be seen from Figure 3 that the most remarkably coherence appears in $13 < tU/h < 32$ (marked by the white rectangle) near the scale $3h$ (marked by the red dash line) and in $tU/h > 35$ at the scale $3h$ to $10h$. Most of the coherence area of $tU/h > 35$ is beyond the white dash line, which is the cone of influence for the wavelet, and the wavelet coherent value is untrusted because of too close to the beginning or ending instantaneous. Thus, we only focus on the $13 < tU/h < 32$ region. The phase angle in this area is about π. This highly coherent area indicates that there are $3h$ scale structures continuously passing the measurement point during $13 < tU/h < 32$. From Figure 2, one can see that there are several hairpin packets in the flow field during $13 < tU/h < 32$. Thus, the highly coherent area near the scale $\lambda/h = 3$ in Figure 3 reveals the passing of several hairpin packets. In addition, the π phase angle further reveals the velocity structure inside the hairpin packets. As shown in Figure 4, when the hairpin packet passes the measurement point, the Q2 events lead to that the streamwise fluctuation u appearing as positive firstly, and the vertical fluctuation v appearing as negative. Both u and v cross 0 when the shear layer passes the measurement point, and then u and v turn into negative and positive, respectively. Therefore, the phase difference between time series u and v has to be approximately 180° when hairpin packets are passing. The π phase angle reveals the feature of the hairpin packets that the Q2 and Q4 events dominate their inside velocity structures.

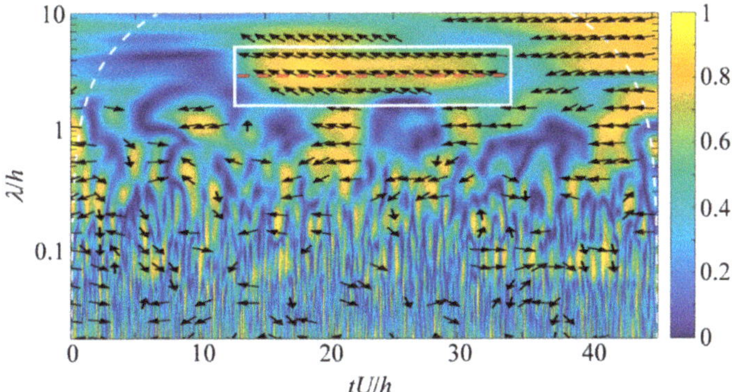

Figure 3. Wavelet coherent coefficients between the streamwise and vertical velocity series at the measurement point ($x/h = 0$, $y/h = 0.5$). The white box marks the time duration as the same as that in Figure 2, and the red dash line indicates the $3h$ scale. The high coherent area appears in the white box and the phase difference is approximately π.

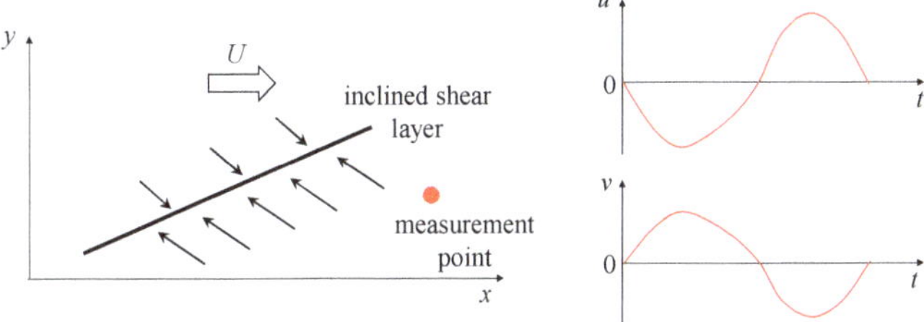

Figure 4. The explanation of the phase angle in the wavelet coherency between the streamwise and vertical velocity series at the same measurement point when the incline structures pass. When the hairpin packet passes the measurement point, the Q2 events lead to that the streamwise fluctuation u appears in positive firstly, and the vertical fluctuation v appears in negative. After the inclined shear layer passes the measurement point, u and v turn into negative and positive, respectively.

The u series at ($x/h = 0$, $y/h = 0.5$) was chosen as the fixed point to do the wavelet coherency with other three measurement points. Figure 5 shows the wavelet coherency of u series at ($x/h = -0.35$, $y/h = 0.15$), ($x/h = 0$, $y/h = 0.15$) and ($x/h = 0.35$, $y/h = 0.15$), respectively. Points ($x/h = -0.35$, $y/h = 0.15$) and ($x/h = 0.35$, $y/h = 0.15$) are located at the upstream and downstream of point ($x/h = 0$, $y/h = 0.5$), respectively, and ($x/h = 0$, $y/h = 0.15$) is directly below the fixed point. From Figure 5a, there are two highly coherent areas during $11 < tU/h = 35$ near the scale $3h$. The first one is roughly from $tU/h = 11$ to 23 (marked by the white box). The second one is approximately from $tU/h = 27$ to 30 and much weaker than the correlation during the same duration in Figure 3, which can be attributed to the fact that the correlation reduces when the distance between two measurement points increases. Referring to Figure 2, these two highly coherent areas are related to the typical hairpin packets during $11 < tU/h < 21$ and $27 < tU/h < 29$, respectively.

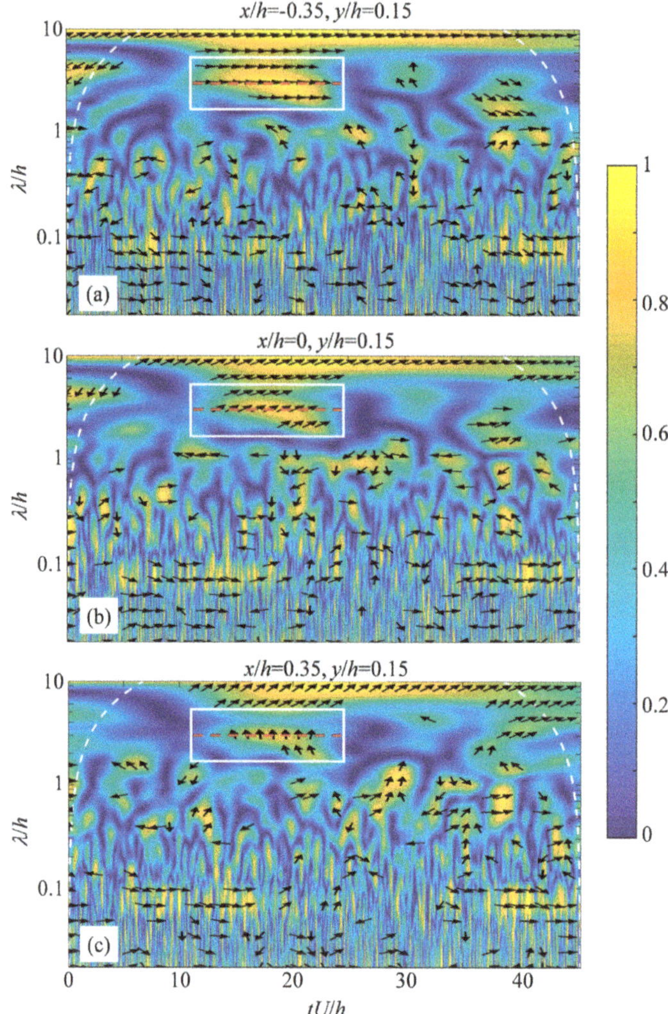

Figure 5. Wavelet coherent coefficients between the streamwise velocity series at different measurement points. The fixed point is ($x/h = 0$, $y/h = 0.5$), the moving points are ($x/h = -0.35$, $y/h = 0.15$), ($x/h = 0$, $y/h = 0.15$), and ($x/h = 0.35$, $y/h = 0.15$). The highly coherent area reduces and the phase angle increases when the lower point moving from the upstream to downstream of the fixed upper point.

The comparison of the coherent area during $11 < tU/h < 21$ between three points in Figure 5a–c shows the coherent area reduces when the measurement point moves from the upstream to downstream. It means the overlap time duration of the $3h$ scale motions between fixed and moving points reduces. This is caused by the incline feature of the $3h$ scale motions. Figure 6 sketches cartoons to show the explanation. Two solid red dots represent two measurement points. Blue and yellow shapes stand for the inclined $3h$ scale structures. The blue and yellow shape mark the starting and end time instant, respectively, that both measurement points are inside the structure. D is the distance between the centers of the blue and yellow shapes. Thus D/U is the time interval of both measurement points located in the structure, which is the highly coherent area in Figure 5. It can be seen from Figure 6 that D is the greatest when the lower point is located at the upstream of the upper point (corresponding to

Figure 5a). *D* decreases when the lower point moves from the upstream to downstream, as shown in Figure 5b,c, because of the incline feature of the hairpin packets. In addition, the phase angle increases when the lower point moves from the upstream to downstream. This is also cased by the incline feature of the hairpin packets. When the lower point is located at the proper location of the upstream of the upper point, the two points almost enter the structure at the same time, as shown in Figure 6a. Thus, the typical signal of the hairpin packet presents at the same time in the velocity series in the two points and the phase difference is small. When the two points are on the vertical line, as shown in Figure 6b, the lower point enters the structure after the upper one enters the structure for a while. Therefore, there is hysteresis between the phases of the signal of the corresponding scale. As the lower point keeps on moving downstream, the phase difference will increase.

From the above discussion, the wavelet coherency analysis between the streamwise and wall-normal velocity series at the same point and the streamwise velocity series from different points not only detects the occurrence and scales of hairpin packets in open channel flows, but also reveal the internal velocity organizations and the tilt geometry of hairpin packets. The wavelet coherent coefficient and the phase angles present a powerful tool for the detection of the energetic coherent structures and the analysis of their internal organizations.

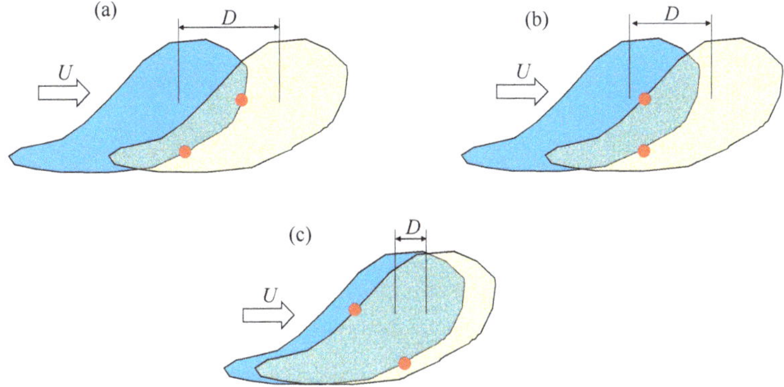

Figure 6. The explanation of the reducing wavelet coherent area and the increasing phase angle when the lower point moving from the upstream to downstream of the fixed upper point. When the lower point is located at the proper location of the upstream of the upper point, the two points almost enter the structure at the same time, as shown in (**a**). Thus, the typical signal of the hairpin packet presents at the same time in the velocity series in the two points and the phase difference is small, as shown in Figure 5a. When the two points are on the vertical line as shown in (**b**), the lower point enters the structure after the upper one enters the structure for a while. There is hysteresis between the phases of the signal of the corresponding scale (see Figure 5b). As the lower point keeps on moving downstream, the phase difference will increase, as in (**c**).

5. Concluding Remarks

Many studies based on single-point measurement data have demonstrated the impressive ability of the wavelet coherency analysis to catch the coherent structures in the wall-bounded flows, however, the question that how the events found by the wavelet coherency analysis based on single point measurement data relate to the features of the three-dimensional coherent structures remains open. Present study employed the TR-PIV system to produce a time series of a two dimensional flow field in the streamwise-wall-normal plane of steady open channel flows, which makes it possible to find out entire flow structures from the 2D flow field of the corresponding wavelet coherent events. Wavelet coherency analysis was applied on the velocity series from a single or multi-point to find the highly energetic and coherent events. The fluctuating velocity fields during the time period with high energy and high wavelet coherence were pieced together based on Taylor's frozen turbulence hypothesis to explore the corresponding entire coherent structures of the wavelet events. The major finding of this study are as follows:

(1) The high value peaks in the pre-multiplied wavelet power spectrum curves stand for the energetic scales in the signal, and the high value areas in the local wavelet spectrum gives both the scales and the time instants of energetic motions. The $3h$ motion is the main energetic scale in open channel flows based on the wavelet spectral analysis on the TR-PIV data.

(2) The high wavelet coherent coefficients area from the wavelet coherency analysis can also detect the scale and the occurrence time instants of energetic motions, and the phase angles reveal the inner structure of the energetic motions.

(3) The wavelet coherency phase angle of $3h$ motions between the streamwise and vertical velocity of the same measurement point is π, which reveals that the velocity structure of these $3h$ motions is organized by the Q2 and Q4 events. By the wavelet coherency analysis between the streamwise velocities from different measurements points, the highly coherent area reduces and the phase angle increases when the lower point moving from the upstream to downstream of the fixed upper point, which reflects the $3h$ scale motions, is an incline structure.

(4) All the features of this $3h$ motions revealed by the wavelet coherency analysis support the hairpin packets model in open channel flows.

Author Contributions: Conceptualization, Q.Z.; methodology, K.C. and Y.Z; data curation, Y.Z.; writing—original draft preparation, K.C.; Y.Z., writing—review and editing, Q.Z.; supervision, Q.Z.; funding acquisition, Q.Z.

Funding: The study was supported by the National Natural Science Foundation of China (Grant No.51809268) and the Fundamental Research Funds for the Central Universities (China Agricultural University, Project No. 2019TC043).

Conflicts of Interest: The authors declare no conflict of interest.

Symbol Table

Symbol	Meaning	Symbol	Meaning
u	streamwise fluctuation	v	vertical fluctuation
U	mean velocity	ψ	wavelet function
W_u	wavelet coefficient	W_{uv}	cross wavelet coefficient
E_{uu}	spectrum	R_{uv}	correlation coefficient
f	Fourier frequency	k	wave number
λ	wavelength	ω_0	wavelet center frequency
$\Re[\cdot]$	real part of complex variables	$\Im[\cdot]$	imaginary part of the complex variables
J	bed slope	h	water depth
u_*	friction velocity	Fr	Froude number
Re	Reynolds number	Re_τ	friction Reynolds number

References

1. Farge, M. Wavelet transforms and their applications to turbulence. *Annu. Rev. Fluid Mech.* **1992**, *24*, 395–458. [CrossRef]
2. Liu, P.C. Wavelet spectrum analysis and ocean wind waves. *Wavel. Geophys.* **1994**, *4*, 151–166.
3. Torrence, C.; Compo, G.P. A practical guide to wavelet analysis. *Bull. Am. Meteorol. Soc.* **1998**, *79*, 61–78. [CrossRef]
4. Torrence, C.; Webster, P.J. Interdecadal changes in the ENSO-monsoon system. *J. Clim.* **1999**, *12*, 2679–2690. [CrossRef]
5. Grinsted, A.; Moore, J.C.; Jevrejeva, S. Application of the cross wavelet transform and wavelet coherence to geophysical time series. *Nonlinear Process. Geophys.* **2004**, *11*, 561–566. [CrossRef]
6. Camussi, R.; Robert, G.; Jacob, M.C. Cross-wavelet analysis of wall pressure fluctuations beneath incompressible turbulent boundary layers. *J. Fluid Mech.* **2008**, *617*, 11–30. [CrossRef]

7. Chui, C.K. *An Introduction to Wavelet*; Academic Press: Cambridge, MA, USA, 2014; Volume 1.

8. Tropea, C.; Yarin, A.L.; Foss, J.F. *Springer Handbook of Experimental Fluid Mechanics*; Springer Science & Business Media: New York, NY, USA, 2007.

9. Nezu, I. Open-channel flow turbulence and its research prospect in the 21st century. *J. Hydraul. Eng.* **2005**, *131*, 229–246. [CrossRef]

10. Nikora, V.; Roy, A.G. Secondary flows in rivers: Theoretical framework, recent advances, and current challenges. *Gravel Bed Rivers Process. Tools Environ.* **2012**, 3–22. [CrossRef]

11. Kähler, C.J.; Tommaso, A.; Vlachos, P.P.; Sakakibara, J.; Hain, R.; Discetti, S.; la Foy, R.; Cierpka, C. Main results of the 4th International PIV Challenge. *Exp. Fluids* **2016**, *57*, 1–71. [CrossRef]

12. Zhong, Q.; Li, D.; Chen, Q.; Wang, X. Coherent structures and their interactions in smooth open channel flows. *Environ. Fluid Mech.* **2015**, *15*, 653–672. [CrossRef]

13. Zhong, Q.; Chen, Q.; Wang, H.; Li, D.; Wang, X. Statistical analysis of turbulent super-streamwise vortices based on observations of streaky structures near the free surface in the smooth open channel flow. *Water Resour. Res.* **2016**, *52*, 3563–3578. [CrossRef]

14. Perrier, V.; Philipovitch, T.; Basdevant, C. Wavelet spectra compared to Fourier spectra. *J. Math. Phys.* **1995**, *36*, 1506–1519. [CrossRef]

15. Guala, M.; Hommema, S.; Adrian, R. Large-scale and very-large-scale motions in turbulent pipe flow. *J. Fluid Mech.* **2006**, *554*, 521–542. [CrossRef]

16. Balakumar, B.; Adrian, R. Large-and very-large-scale motions in channel and boundary-layer flows. *Philos. Trans. R. Soc. A Math. Phys. Eng. Sci.* **2007**, *365*, 665–681. [CrossRef] [PubMed]

17. Monty, J.P.; Hutchins, N.; Ng, H.C.H.; Marusic, I.; Chong, M.S. A comparison of turbulent pipe, channel and boundary layer flows. *J. Fluid Mech.* **2009**, *632*, 431–442. [CrossRef]

18. Zhou, J.; Adrian, R.J.; Balachandar, S.; Kendall, T.M. Mechanisms for generating coherent packets of hairpin vortices in channel flow. *J. Fluid Mech.* **1999**, *387*, 353–396. [CrossRef]

19. Adrian, R.J. Hairpin vortex organization in wall turbulence. *Phys. Fluids* **2007**, *19*, 041301. [CrossRef]

20. Nezu, I.; Sanjou, M. PIV and PTV measurements in hydro-sciences with focus on turbulent open-channel flows. *J. Hydro Environ. Res.* **2011**, *5*, 215–230. [CrossRef]

21. Adrian, R.; Meinhart, C.; Tomkins, C. Vortex organization in the outer region of the turbulent boundary layer. *J. Fluid Mech.* **2000**, *422*, 1–54. [CrossRef]

22. Adrian, R.J.; Marusic, I. Coherent structures in flow over hydraulic engineering surfaces. *J. Hydraul. Res.* **2012**, *50*, 451–464. [CrossRef]

Article

Bedform Morphology in the Area of the Confluence of the Negro and Solimões-Amazon Rivers, Brazil

Carlo Gualtieri [1,*], Ivo Martone [1], Naziano Pantoja Filizola Junior [2] and Marco Ianniruberto [3]

[1] Department of Civil, Architectural and Environmental Engineering (DICEA), University of Naples Federico II, 80125 Naples, Italy; ivo.martone@unina.it
[2] Department of Geosciences, Federal University of Amazonas, Manaus CEP 69077-000, Brazil; nazianofilizola@ufam.edu.br
[3] Institute of Geosciences, University of Brasilia, Brasilia CEP 70910-900, Brazil; ianniruberto@unb.br
* Correspondence: carlo.gualtieri@unina.it

Received: 6 May 2020; Accepted: 3 June 2020; Published: 6 June 2020

Abstract: Confluences are common components of all riverine systems, characterized by converging flow streamlines and the mixing of separate flows. The fluid dynamics of confluences possesses a highly complex structure with several common types of flow features observed. A field study was recently conducted in the area of the confluence of the Negro and Solimões/Amazon Rivers, Brazil, collecting a series of Acoustic Doppler Current Profiler (ADCP) transects in different flow conditions. These data were used to investigate the morphology of the bedforms observed in that area. First, the bedforms were mostly classified as large and very large dunes according to Ashley et al. (1990), with an observed maximum wavelength and wave height of 350 and 12 m, respectively. Second, a comparison between low flow and relatively high flow conditions showed that wavelength and wave height increased as the river discharge increased in agreement with previous literature studies. Third, the lee side angle was consistently below 10°, with an average value of about 3.0°, without flow separation confirming past findings on low-angle dunes. Finally, a comparison between the bedform sizes and past literature studies on large rivers suggested that while several dunes were in equilibrium with the flow, several largest bedforms were found to be probably adapting to discharge changes in the river.

Keywords: river hydrodynamics; ADCP; bedforms morphology; river confluence; Amazon River

---

1. Introduction

Bedforms are a very common feature in alluvial fluvial channels as a result of the unstable interaction between water flow, sediment transport and bed morphology [1,2]. Bedform analyses range from fundamental analytical descriptions [3], to detailed numerical simulations [4–6] and laboratory measurements [7–11], to large-scale field measurements [12–15]. In the engineering context, these analyses are fundamental in predicting discharge, flow resistance [16] and sediment transport in rivers. In geology, bedforms are studied as primary sedimentary structures that are forming at the time of deposition of the sediment and reflect the characteristics of the depositional environment [17,18].

Bedforms are often approximated by triangular shapes, but natural bedforms present a more complicated bi and three-dimensional morphology. The broadest classification of unidirectional flow bedforms is based on the flow regime under which the bedforms develop. Simons and Richardson [19] distinguished a lower flow regime, for subcritical flows, associated with ripples and dunes, and through a transitional flat bed, an upper flow regime, for supercritical flows, with standing waves and antidunes. The upper flow regime is uncommon in deep rivers, where the observed bedforms are mainly ripples and dunes [13,20]. Several criteria were proposed to identify ripples and dunes. Ripples range in length from approximately 0.05 to about 0.6 m and in height from 0.005 to just less than 0.05 m [17], but, according to Yalin [21], ripples have length and height less than 0.6 and 0.04 m, respectively. Ashley et al. [22] classified dunes into four groups based on height (H_{bf}) and wavelength (λ_{bf}), as listed in Table 1.

Table 1. Classification of bedforms according to Ashley et al. [22].

Main Group	Group	H_{bf} (m)	λ_{bf} (m)
Ripples	Ripples (or small dunes)	0.075–0.40	0.6–5.0
	Medium dunes	0.40–0.75	5.0–10.0
Dunes	Large dunes	0.75–5.0	10.0–100.0
	Very large dunes	>5.0	>100.0

While ripple size is scaling with the size of the grains of the bed and it is independent of the flow depth [23], dunes size is related to flow conditions. Several equations were proposed in the literature to predict dune sizes and their adjustment to flow changes [9–11,13,23–26]. Dune development is controlled by the interaction among the flow, sediment transport, and dune form. Deformation and adaptation have been recognised as the main mechanisms responsible for dune development [9]. Even where the statistical descriptors of the dune population have converged (equilibrium state) and the reach-averaged bed shear stress is constant, variability in dune shape is caused by continuously deformation as they migrate. Additional variability is introduced by the adaptation of dunes to changes in flow. Dune adaptation is ubiquitous because river flow is typically both unsteady and non-uniform at the temporal and spatial scales that are needed for dunes equilibrium [9]. However, equilibrium conditions are not frequent because they require time and a sufficient rate of sediment transport. Bedforms increasing/decreasing in size over the time are called developing/diminishing dunes, while they may be considered in equilibrium if they migrate without any change in shape or mass, i.e., deformation [20]. The temporal lag of the development of dunes relative to their formative flow, i.e., dune hysteresis [9], directly depends on their size [13,27]. Dunes are generally seen to grow in size during the rising stage, reaching their maximum development after peak discharge. However, it was found that water depth and flow velocity have separate effects on dune adaptation. Dunes crests/troughs do not respond simultaneously to changes in flow and crest flattening, i.e., increasing bedform length while height is decreasing, is related to decreasing depth and increasing flow velocity [9]. During the falling stage, dunes may stretch and flatten with a rapid downstream migration of the lee side, remaining as large bedforms during the subsequent low-flow period [9,13,20].

According to Best [1], the fluid dynamics of asymmetric river dunes with an angle-of-repose lee side and generated in a steady, uniform unidirectional flow, is characterised by (1) accelerating flow over the dune stoss side; (2) flow separation or deceleration in the lee of the dune, with reattachment at from 4 to 6 dune heights downstream; (3) a shear layer bounding the separation zone, which divides this recirculating flow from the free stream fluid above; (4) an expanding flow region in the dune lee side; and (5) downstream of the reattachment point, a new boundary layer that grows beneath the wake along the stoss slope of the next dune downstream.

Such flow structure has many important implications for flow resistance [1]. The differential pressures generated by flow separation and flow acceleration/deceleration associated with the dune form generate a net force on the dune, called the 'form drag', which, together with the grain roughness drag, called the "skin drag", determines dune morphology and flow resistance [28,29]. However, low-angle dunes (lee-side angle < 10°) do not possess a zone of permanent flow separation, and those with lee-side angles < 4° are believed to possess no flow separation at all, resulting in lower energy losses [15]. In this regard, a recent analysis of high-resolution bathymetry data demonstrated that the largest rivers on Earth are characterized by low-angle lee-sides (mean ~10°) [15].

Dune features are expected to show an even greater complexity at river confluences, as these are characterized by very complex hydrodynamics and morphodynamics located in the Confluence Hydrodynamic Zone (CHZ) [30,31]. The CHZ is often characterised as a stagnation zone, a velocity deflection and re-alignment zone, a separation region with recirculation, a maximum velocity and flow recovery region [32]. The central part of the CHZ ends where flow recovery starts, while the CHZ ends where the flow is no more significantly affected by the confluence. The hydrodynamics and morphodynamics within the CHZ are influenced by: the planform of the confluence; the momentum flux ratio of merging streams; the level of concordance between channel beds at the confluence entrance; and differences in the water characteristics (temperature, conductivity, suspended sediment concentration) between the incoming tributary flows lead to the development of a mixing interface and may impact local processes about the confluence [32,33]. Confluence bed morphology is characterized by the presence of a scour hole, bars (tributary-mouth, mid-channel and bank-attached bars), and a region of sediment accumulation near the upstream junction corner [30].

This paper presents and analyses the morphology of the bedforms observed during two field surveys carried out at the confluence between Rio Negro and Rio Solimões in the Amazon Basin. The surveys were carried out using Acoustic Doppler Current Profiler (ADCP) during low flow in 2014 and in relatively high flow in 2015. The objectives of this study are to (1) describe the bedforms characteristics at the Negro/Solimões confluence; (2) compare bedform characteristics in different flow conditions; and (3) compare bedform scales with those derived from literature theoretical/empirical equations and from past field studies conducted in large rivers.

2. Field Site and Campaigns. Basic Results on Hydrodynamics and Morphodynamics

2.1. Hydrological and Sedimentological Background

The study area is centered about the confluence of the Negro and Solimões Rivers, located near Manaus in Northern Brazil, where these rivers merge to form the Amazon River, (Figure 1). The Negro and Solimões confluence ranks among the largest on Earth and is famous for visibly revealing the meeting of the black (Negro) and white (Solimões) waters of the two rivers.

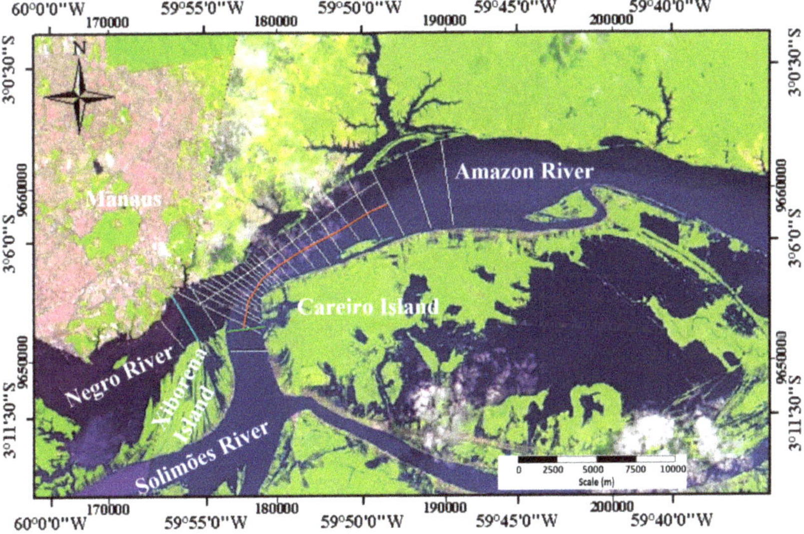

Figure 1. Map of area of the Negro/Solimões confluence. White and coloured solid lines represent transects surveyed with Acoustic Doppler Current Profiler (ADCP) during FS-CNS1 and FS-CNS2 (Long1: red line; N-CNS: blue line; S-CNS: green line). Background: Landsat 8 image. Map datum: WGS84; projection: UTM 21S.

The distinct waters of these two rivers are related to the geology (soils, soil coverage, etc.) of the two respective catchments within the Amazon Basin. The Negro sub-basin is located in the North draining the western slopes of the Guyana Shield, which is characterised by gentle gradients and densely vegetated margin, which, in turn, implies a low sediment production [34], while the Solimões catchment includes the eastern margin of the Andes, where the combination of high declivity and erodible rocks gives origin to high sediment production [35,36]. The mean water discharge of the Negro and Solimões Rivers is about 30,000 and 100,000 m^3/s, accounting for 14% and 49%, respectively, of the total freshwater discharge of the Amazon River into the Atlantic Ocean [34]. The two rivers have a different hydrological cycle: the Negro has two distinct discharge peaks along the year, the first of low-amplitude during the first three months of the year, and the second larger in the middle of the year; the Solimões has one peak between May and June [34]. Hydrologic data collected at the fluviometric stations, located in Tatu-Paricatuba (Negro River) and Manacapuru (Solimões River) (Figure 2), (www.ore-hybam.org/) show that low-flow conditions are occurring during the local Autumn, while the high-flow conditions occur during the local Winter season (June–July) (Figure 3). In terms of sediment load, the difference between the two rivers is even larger. The Solimões River has at the Manacapuru station an average load of suspended solids of 14,174 Kg/s, accounting for more than half of the total load of the Amazon River into the Atlantic Ocean, while the Negro River has at Paricatuba station an average suspended load of 254 Kg/s [37].

Figure 2. Location of the Tatu-Paricatuba and Manacapuru stations.

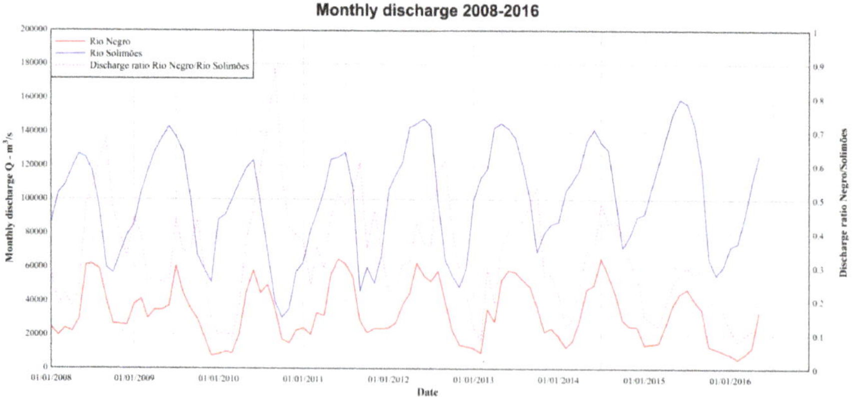

Figure 3. Discharge time series from 2008 to 2016.

2.2. Field Campaigns

The field campaigns were carried out about the Negro/Solimões confluence in the Amazon River basin within the EU-funded Clim-Amazon Project, to study low flow (October 2014, FS–CNS1 campaign) and relatively high flow (April/May 2015, FS–CNS2 campaign) conditions [38–43]. The discharges measured during the campaigns were within the typical observed values for the seasons and daily differences were at most 5%. The surveys were carried out using an Acoustic Doppler Current Profiler (ADCP) as well as a multi-parameter probe for the measurement of water physico-chemical parameters (temperature, conductivity, turbidity, etc.) and total suspended sediment (TSS) concentration. During the surveys, a Teledyne RDI 600 kHz Rio Grande ADCP was used to collect flow velocity, water column backscattering [44] and water depth at key locations about the confluence. In total, 98 cross-sectional transects were collected. In addition, 2 and 3 longitudinal profiles along both sides of the Amazon River were collected in FS–CNS1 and FS–CNS2 surveys, respectively.

2.3. Basic Observations of Hydrodynamics and Morphodynamics about the Confluence

Table 2 lists the main flow properties of the Negro and Solimões rivers measured just upstream of the confluence (N-CNS and S-CNS, three for each river and each field campaign) during the two surveys. Large differences in discharge and flow velocities were observed in the Solimões River between the two surveys, whereas, on the Negro River, these differences were smaller.

Table 2. Main flow properties of Negro and Solimões Rivers during FS–CNS1/FS–CNS2/FS–CNS3. Q = median discharge; A = median cross-sectional area; W = channel median width; h_{med} = median depth; W/h_{rect} = median of the aspect ratio; V_{avg} = median of the cross-section velocity (Q/A); $V_{depth\text{-}avg}$ = median of the depth-averaged velocity; Dir = median of flow direction degrees from North; V_{max} = maximum depth-averaged velocity.

River	Field Trip	Q (m³/s)	A (m²)	W (m)	h_{med} (m)	W/h_{rect} (-)	V_{avg} (m/s)	$V_{depth\text{-}avg}$ (m/s)	Dir (°)	V_{max} (m/s)
Negro	FS–CNS1	24,510	64,784	2830	24.4	117	0.38	0.39	59	0.69
	FS–CNS2	33,501	86,952	2875	31.2	95	0.38	0.40	58	0.67
Solimões	FS–CNS1	63,380	42,789	1589	27.2	59	1.49	1.33	289	2.20
	FS–CNS2	105,205	61,895	1925	28.6	60	1.70	1.52	255	2.59

From FS–CNS1 to FS–CNS2, the maximum depth-averaged velocity was almost constant in the Negro River, but the Solimões River increased from 2.2 to 2.6 m/s. Furthermore, from low to high flow conditions, the Negro channel increased in depth, from 24 to 31 m, but not in width, whereas in the Solimões River the width increased from 1.6 to 1.9 km and the depth from 27 to 28 m. Finally, from FS–CNS1 to FS–CNS2, the median flow direction in the Negro River remained unchanged, whereas in the Solimões River a significant change in direction occurred. It is worth noting that the confluence junction angle is of about 65° (Figure 1). At the confluence entrance, the Negro channel is almost aligned with the Amazon channel, while the Solimões-Amazon waters must undergo a large change in flow direction (60°–70°) to enter in the Amazon channel. Common hydrodynamic features [32] already noted in past confluence studies were present even about the Negro/Solimões confluence. The approximate location of those features is shown in Figure 4, where they are numbered as: (1) the stagnation zone; (2) the region of deflection; (3) the region of maximum velocity; (4) the downstream separation zone; (5) the beginning of the region of flow recovery; and (6) the end of the CHZ. Further details about hydrodynamics are described in [41].

Ianniruberto et al. [43] identified that: upstream of the confluence, the Negro side is mostly characterized by a rocky bed with fine sand cover, whilst on the Solimões side the river bed consists predominantly of sand, and a sediment deposition region occurs at the junction corner in correspondence with the stagnation zone (Figure 4, feature 1); the central part of the CHZ is characterised by a sediment by-pass region exposing eroded Cretaceous bedrock, with a scour hole corresponding to the region of maximum velocity (Figure 4, feature 3); deposited sediment forming a bank-attached bar on the Solimões side of the CHZ (Figure 4, feature 4); avalanche faces of sediments at the Solimoes mouth;

a bedrock terrace was observed towards the downstream end of the CHZ (Figure 4, feature 5), marking the transition from rocky to alluvial bed, where the bedforms were found.

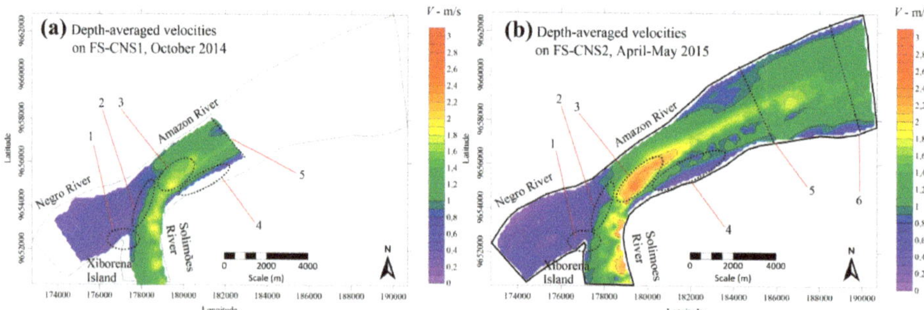

Figure 4. Map of the depth-averaged velocities about Negro/Solimões confluence on (**a**) FS–CNS1 and (**b**) FS–CNS2, with the location of the hydrodynamics features. Legend: (1) stagnation zone; (2) region of deflection; (3) region of maximum velocity; (4) downstream separation zone; (5) beginning of the region of flow recovery; and (6) end of the confluence hydrodynamic zone.

3. Bedforms morphology. Results and Discussion

3.1. Hydrodynamics and Sediment Transport Parameters along the Longitudinal Transects. Results

The location and length of the two longitudinal transects collected in the area of the confluence during FS–CNS1 and FS–CNS2, approximately 900 m away from the Solimões/Amazon right bank, are presented in Table 3 and Figure 1.

Table 3. Locations and length of the ADCP transects collected in FS–CNS1 and FS–CNS2.

Transect Name	Start (Lat, Long)°	End (Lat, Long)°	Length (m)
Long1_11_14_000	−3.141315, −59.89666	−3.083647, −59.828362	9843
Long1_03_05_15_000	−3.147089, −59.895264	−3.071548, −59.78756	14,133

The ADCP measurements provided data about water depth and velocity. The observed water depth ranged from 21 to 68 m, with an increase of about 6–7 m from low to relatively high flow conditions. Table 4 lists the minimum, average, median and maximum value of the depth-averaged velocity along the ADCP transects as well as its standard deviation. No large variations were observed from low flow to relatively high flow conditions on minimum and average velocity, but the maximum depth-averaged velocity increased from 2.2 to 2.8 m/s. Figures 5 and 6 show the longitudinal distribution of the depth-averaged velocity in the ADCP transects collected in FS–CNS1 and FS–CNS2, respectively.

Table 4. Depth-averaged velocity in the ADCP transects collected in FS–CNS1 and FS–CNS2. Legend: V_{min} = minimum depth-averaged velocity; V_{mean} = mean of the depth-averaged velocity; $V_{st.dev.}$ = standard deviation of the depth-averaged velocity; $V_{depth-avg}$ = median of the depth-averaged velocity; V_{max} = maximum depth-averaged velocity.

Transect	V_{min} (m/s)	V_{mean} (m/s)	$V_{st.dev.}$ (m/s)	$V_{depth-avg}$ (m/s)	V_{max} (m/s)
Long1_11_14_000	0.83	1.49	0.17	1.47	2.16
Long1_03_05_15_000	0.96	1.71	0.35	1.63	2.84

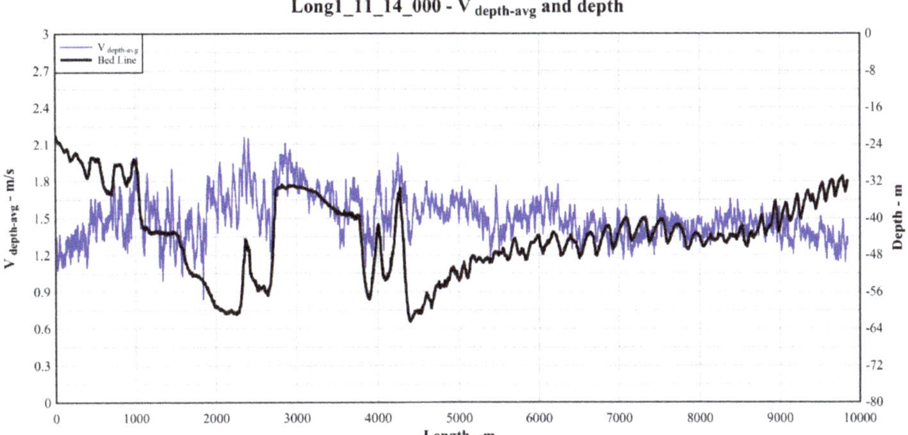

Figure 5. Depth-averaged velocity (blue) and bedline (black) for Long1_11_14_000 transect (FS–CNS1).

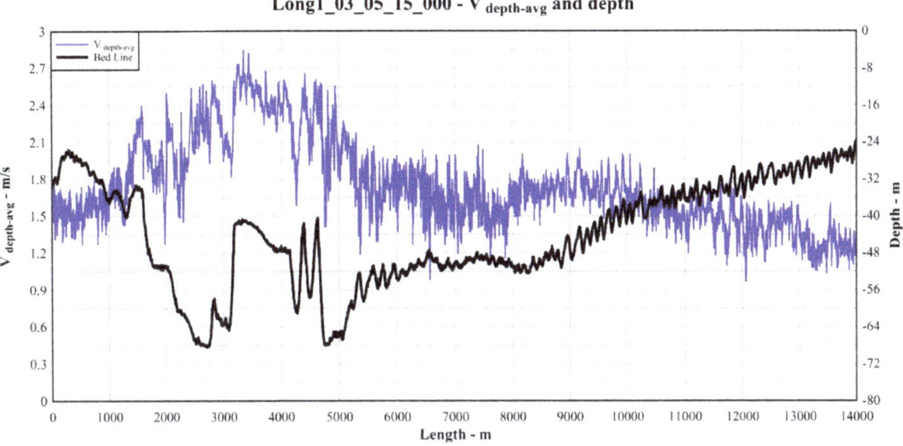

Figure 6. Depth-averaged velocity (blue) and bedline (black) for Long1_03_05_15_000 transect (FS–CNS2).

Downstream from the mouth of the Solimões, the longitudinal transect was initially located on the rocky scour hole, with a depth larger than 60 m. After the scour hole, it was observed a wide sandstone terrace at depth of about 35 m, gently sloping downstream, followed by sharp 30 m decrease in depth. Downstream of this large depression, the bedforms were found [43]. The starting point of the bedforms is located within the flow recovery region, where the Amazon channel is widening, at approximately 4.7 km downstream of the confluence junction. Interestingly, the bedform size seems to increase and the shape changes with downstream distance.

Some parameters related to sediment transport were calculated. In alluvial channels, friction is related both to the grain resistance and to the form of bedform and the total shear stress is [2,45]:

$$\tau_b = \tau_b' + \tau_b''$$ (1)

where τ'_b and τ''_b are the skin friction shear stress and the form-related shear stress, respectively.

The skin friction shear stress was calculated as [46]:

$$\tau_b' = \rho C_d V_{depth-avg}^2 \tag{2}$$

where C_d is the drag coefficient, which was obtained as:

$$C_d = \frac{\kappa^2}{\ln^2(h/e\,z_0)} \tag{3}$$

where κ is the von Kármán constant, e is the Euler number and z_0 is the zero−velocity height above the bed, which can be obtained as [47]:

$$z_0 = 0.1\,d_{84} \tag{4}$$

where d_{84} is bed grain diameter such that 84% of diameters are finer (Figure 7) [48]. The water density was calculated from the temperature T as:

$$\rho = -0.0054\,T^2 + 0.021\,T + 1000\ (\text{kg/m}^3) \tag{5}$$

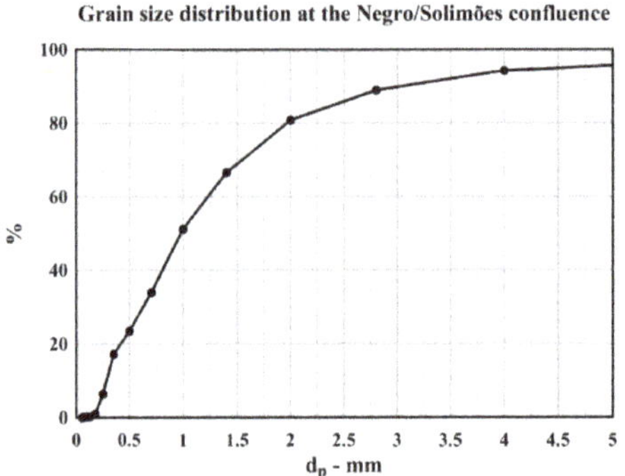

Figure 7. Grain size distribution [48].

The form-related shear stress was computed as [2]:

$$\tau_b'' = \frac{1}{2}\rho V_{depth-avg}^2 \frac{H_{bf}^2}{L_{bf} h} \tag{6}$$

where H_{bf} and L_{bf} are bedforms height and length, respectively, and h is water depth.

Table 5 lists the minimum, average, standard deviation, median and maximum value of the bed shear stress along the ADCP transects calculated using Equation (1). No large variations were observed in the median value from low to relatively high flow conditions. Finally, the shear stress allows us to calculate the maximum suspended grain size d_{ss} [49]:

$$d_{ss} = \sqrt{\frac{18\rho v \cdot 0.8 \cdot \sqrt{\tau_b/\rho}}{g(\rho_s - \rho)}}\ (m) \tag{7}$$

where ρ_s is the particle density, and v is the water kinematic viscosity, that was calculated as:

$$v = (5.85\ 10{-}10 * T^2) - (4.85\ 10{-}8 * T + 1.74\ 10{-}6)\ (\text{m}^2/\text{s}) \tag{8}$$

Table 5. Bed shear stress τ_b. Legend: $\tau_{b\text{-}min}$ = minimum bed shear stress; $\tau_{b\text{-}mean}$ = mean of the bed shear stress; $\tau_{b\text{-}median}$ = median of the bed shear stress; $\tau_{b\text{-}max}$ = maximum bed shear stress.

Transect	$\tau_{b\text{-}min}$ (Pa)	$\tau_{b\text{-}mean}$ (Pa)	$\tau_{b\text{-}median}$ (Pa)	$\tau_{b\text{-}max}$ (Pa)
Long1_11_14_000	1.66	4.44	4.11	10.96
Long1_03_05_15_000	2.29	6.52	5.39	14.26

Table 6 lists the minimum, average, standard deviation, median and maximum value of the maximum suspended grain size along the ADCP transects calculated using Equation (7). The maximum suspended grain sizes were generally in the order of fine sand (0.125–0.250 mm) with a highest value in the range of medium sand (0.250–0.500 mm) in both flow conditions.

Table 6. Maximum suspended grain size d_{ss}. Legend: $d_{ss\text{-}min}$ = minimum value of the maximum suspended grain size; $d_{ss\text{-}mean}$ = mean of the maximum suspended grain size; $d_{ss\text{-}st.dev}$ = standard deviation of the maximum suspended grain size; $d_{ss\text{-}median}$ = median of the maximum suspended grain size; $d_{ss\text{-}max}$ = maximum value of the maximum suspended grain size.

Transect	$d_{ss\text{-}min}$ (mm)	$d_{ss\text{-}mean}$ (mm)	$d_{ss\text{-}st.dev}$ (mm)	$d_{ss\text{-}median}$ (mm)	$d_{ss\,max}$ (mm)
Long1_11_14_000	0.181	0.227	0.002	0.227	0.290
Long1_03_05_15_000	0.197	0.245	0.038	0.244	0.448

3.2. Bedform Morphology. Results and Discussion

Bedform characteristics were derived through several steps. The raw ADCP data were first extracted with WinRiver II. Then, the ADCP data were processed to get depth-averaged vertical and streamwise velocities according to the procedure described in Bahmanpouri et al. [50] with the addition of a further low-pass filtering to remove spikes and noise. The next stage was to track the bottom profile to detect the bedforms as anomaly relative to a reference depth. To this aim, a Matlab code was implemented: a seventh-grade polynomial fit curve was chosen (Figure 8) to define a reference depth used to detect using visual analysis individual bedforms as a succession of trough–crest–trough and to estimate their wavelength and wave height. Using this procedure, further metrics such as wave steepness, lee side and stoss side lengths and angles were calculated.

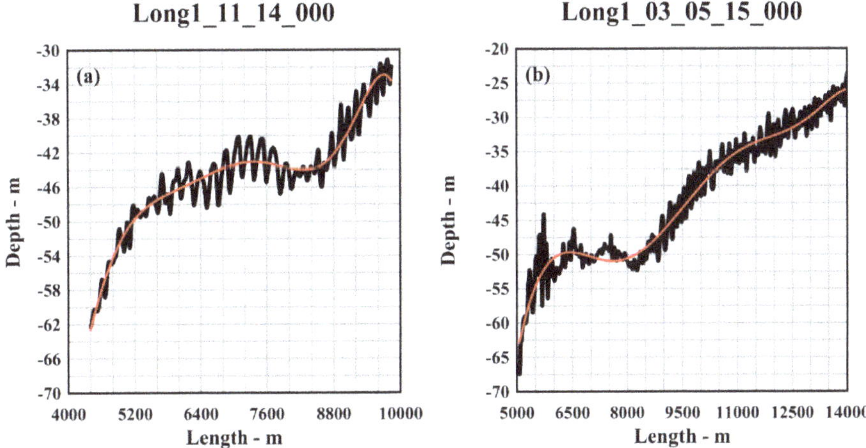

Figure 8. Polynomial seventh grade fit and longitudinal transects for (a) FS–CNS1 and (b) FS–CNS2.

Tables 7–9 list the minimum, average, median and maximum value of the wavelength λ_{bf}, wave height H_{bf} and wave steepness H_{bf}/λ_{bf} as well as their standard deviation for the bedforms observed during this study. The datasets for FS–CNS1 and FS–CNS2 were termed "Encontro das Aguas—1 November 2014" and "Encontro das Aguas—3 May 2015", respectively. Figures 9–11 show the frequency distribution for these parameters. For each parameter 14 classes in size were considered. The 2D analysis identified 36 and 70 bedforms, which were all classified from Table 1 [22] at least as *large dunes* ($\lambda_{bf} > 10$ m), while *very large dunes* ($\lambda_{bf} > 100$ m) were 86% and 70%, in FS–CNS1 and FS–CNS2, respectively. The minimum wavelength λ_{bf} was equal to 55 m, observed in the FS–CNS2. The maximum wavelength was found in relatively high flow conditions and it was longer than 330 m. On average, the wavelength λ_{bf} was 150 and 128 m for FS–CNS1 and FS–CNS2, respectively, while the average wave height H_{bf} was 3.7 m in both cases. The average steepness H_{bf}/λ_{bf} was of 2.5% and 3.0% for FS–CNS1 and FS–CNS2, respectively. The distribution of the wavelength of the observed bedforms had a small percentage (14% and 30%) of bedforms shorter than 100 m.

Table 7. Wavelength λ_{bf}. Legend: $\lambda_{bf\text{-}min}$ = minimum wavelength; $\lambda_{bf\text{-}mean}$ = mean of the wavelength; $\lambda_{bf\text{-}st.dev}$ = standard deviation of the wavelength; $\lambda_{bf\text{-}median}$ = median of the wavelength; $\lambda_{bf\text{-}max}$ = maximum wave length.

Dataset	n.	$\lambda_{bf\text{-}min}$ (m)	$\lambda_{bf\text{-}mean}$ (m)	$\lambda_{bf\text{-}st.dev.}$ (m)	$\lambda_{bf\text{-}median}$ (m)	$\lambda_{bf\text{-}max}$ (m)
Encontro das Aguas—1 November 2014	36	68.12	150.40	50.09	133.30	242.67
Encontro das Aguas—3 May 2015	70	55.35	128.33	49.16	119.15	334.56

Table 8. Dune height H_{bf}. Legend: $H_{bf\text{-}min}$ = minimum wave height; $H_{bf\text{-}mean}$ = mean of the wave height; $H_{bf\text{-}st.dev.}$ = standard deviation of the wave height; $H_{bf\text{-}median}$ = median of the wave height; $H_{bf\text{-}max}$ = maximum wave height.

Dataset.	n.	$H_{bf\text{-}min}$ (m)	$H_{bf\text{-}mean}$ (m)	$H_{bf\text{-}st.dev.}$ (m)	$H_{bf\text{-}median}$ (m)	$H_{bf\text{-}max}$ (m)
Encontro das Aguas—1 November 2014	36	0.78	3.70	1.65	3.78	6.97
Encontro das Aguas—3 May 2015	70	0.76	3.67	2.12	3.72	12.63

Table 9. Dune steepness H_{bf}/λ_{bf} (multiplied by 100). Legend: $(H_{bf}/\lambda_{bf})_{min}$ = minimum wave steepness; $(H_{bf}/\lambda_{bf})_{mean}$ = mean of the wave steepness; $(H_{bf}/\lambda_{bf})_{st.dev.}$ = standard deviation of the wave steepness; $(H_{bf}/\lambda_{bf})_{median}$ = median of the wave steepness; $(H_{bf}/\lambda_{bf})_{max}$ = maximum wave steepness.

Dataset	n.	$(H_{bf}/\lambda_{bf})_{min}$	$(H_{bf}/\lambda_{bf})_{mean}$	$(H_{bf}/\lambda_{bf})_{st.dev.}$	$(H_{bf}/\lambda_{bf})_{median}$	$(H_{bf}/\lambda_{bf})_{max}$
Encontro das Aguas—1 November 2014	36	0.53	2.54	1.09	2.52	4.82
Encontro das Aguas—3 May 2015	70	0.36	3.00	1.50	2.93	11.15

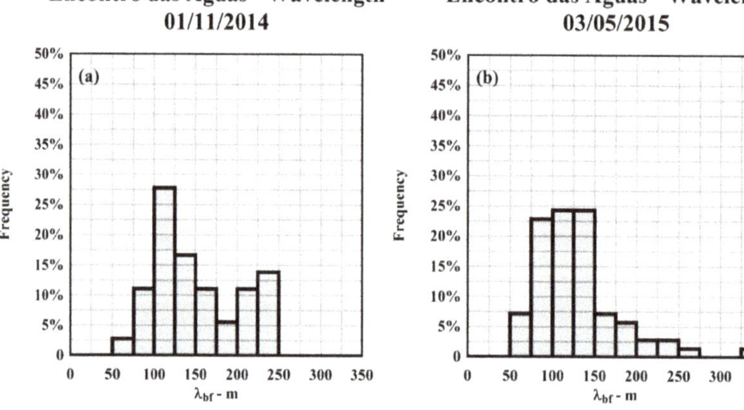

Figure 9. Frequency distribution of bedform wavelength λ_{bf} for (a) FS–CNS1 and (b) FS–CNS2.

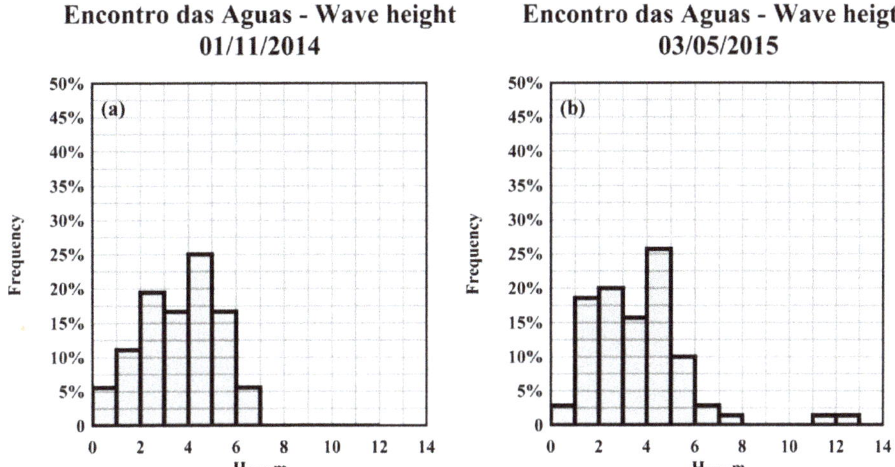

Figure 10. Frequency distribution of bedform wave height H_{bf} for (**a**) FS–CNS1 and (**b**) FS–CNS2.

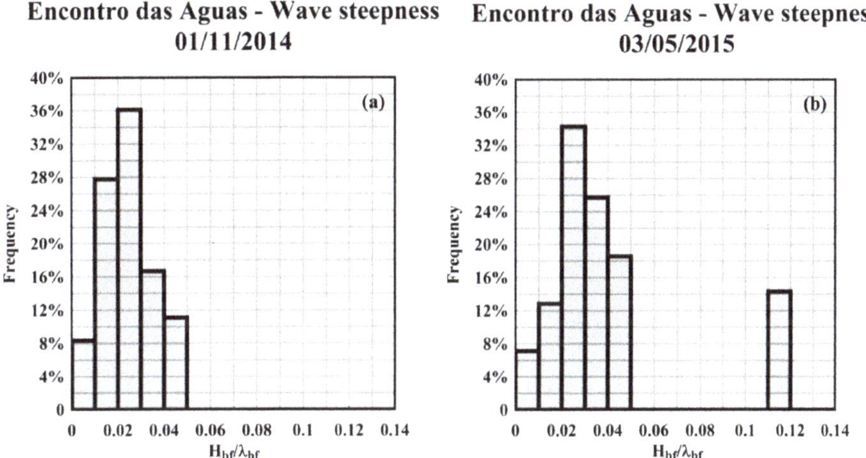

Figure 11. Frequency distribution of bedform steepness H_{bf}/λ_{bf} for (**a**) FS–CNS1 and (**b**) FS–CNS2.

The lee side angle was constantly below 10°, with a maximum value of 8.47° and 8.87° and an average value of 3.02° and 3.25°, in FS–CNS1 and FS–CNS2, respectively. These values are generally consistent with those from large rivers where low-angle (<10°) lee-side slopes are predominant [15]. The asymmetry, defined as the ratio of stoss side length to the bedform length, was on average 0.56 and 0.47 in FS–CNS1 and FS–CNS2, respectively.

In the rising stage from October 2014 (FS–CNS1) to April/May 2015 (FS–CNS2), on average, the wavelength decreased, the steepness increased and the wave height remained unchanged, while the maximum sizes increased. Furthermore, a comparison between the frequency distribution of bedform size in low and relatively high flow conditions showed an increase in wavelength and wave height as the river discharge increased, in agreement with the past literature studies (Figures 9–11). However, as the two ADCP transects are different in length of about 4 km and have a different number of bedforms, the comparison was repeated considering only the bedforms located on the same reach of the two longitudinal transects. The results confirmed the above findings. At the end, the bedforms observed in this study were generally characterised by large wavelengths, ranging from 55 to 335 m, with a median value of 133 and 120 m, respectively, and wave height on average larger than 3 m. The wave steepness was in the range from 0.3% to 11%.

3.3. Modeling Bedforms Morphology. A Comparison with Predictive Equations. Discussion

As already pointed out, dune development is related to both their deformation during migration and their adaptation to flow variations [9]. Dune adaptation has been extensively investigated as an important process in river morphodynamics, but there is not yet a universal model to predict changes in dune sizes in response to flow variations. This morphological response has been often related to sediment mobility, which itself is a product of flow depth and velocity [9], and dune size has been related to flow depth as a result of interaction between large eddies in the flow and the sediment bed [17,51,52], Bedform wavelength was plotted versus bedform wave height [13,20,26,53] and the data were compared with the empirical relationships proposed by Flemming, which were based on 1491 deep sea, tidal and river bedforms [22]:

$$H_{bf} = 0.068 \, \lambda_{bf}^{0.81} \ (\text{m}) \tag{9}$$

$$H_{bf-max} = 0.16 \, \lambda_{bf}^{0.84} \ (\text{m}) \tag{10}$$

where Equation (9) represents a range of steepness H_{bf}/λ_{bf} from 0.08 to 0.1 [20]. Figure 12 shows the results for the bedforms observed in the field surveys, including their respective averages. The data were also compared with the equation proposed by Chen et al. [13] using experimental data collected in the middle–lower Changjiang (Yangtze) River (China):

$$H_{bf} = 0.23 \, \lambda_{bf}^{0.56} \ (\text{m}) \tag{11}$$

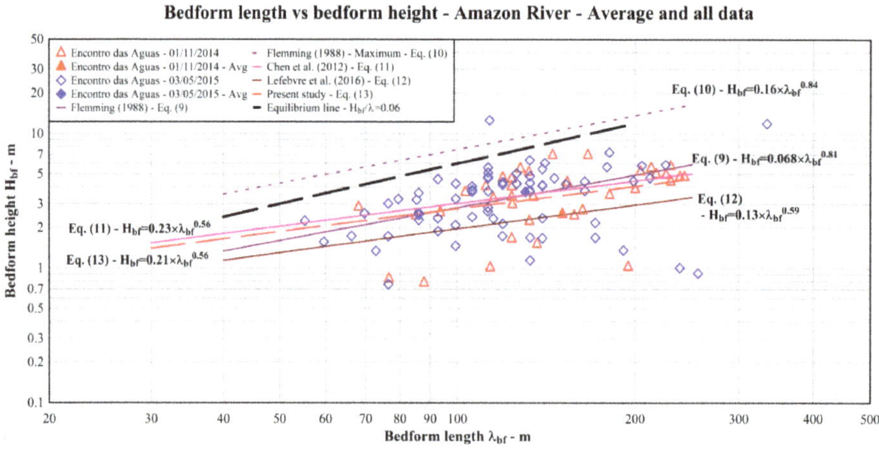

Figure 12. Bedform wavelength vs. bedform wave height.

Furthermore, the equation proposed by Lefebvre et al. [53], who used the data from the Rio Paranà (Argentina) [12] and those from the Lower Rhine (the Netherlands) [54], was included:

$$H_{bf} = 0.13 \, \lambda_{bf}^{0.59} \ (\text{m}) \tag{12}$$

Using the data from the two field surveys, it was possible to derive a new regression equation:

$$H_{bf} = 0.21 \, \lambda_{bf}^{0.56} \, [\text{m}] \tag{13}$$

which is very close to Equation (11). Furthermore, theory and laboratory studies in uniform and steady flow suggest that dunes with a steepness H_{bf}/λ_{bf} less than 0.06 are either non-equilibrium bedforms or represent an equilibrium adjustment of the bed form, in which maximum steepness is precluded by hydraulic constraints, notably a depth limitation [20]. The equilibrium line corresponding to $H_{bf}/\lambda_{bf} = 0.06$ was also included in Figure 12. All the bedforms, but one, were below Flemming's maximum line (Equation (10)). Only some of *large dunes* ($10.0 > \lambda_{bf} > 100$ m) were well aligned between Equation (11) and Equation (9) and close to the equilibrium line, while in many cases they showed a large scatter from these lines. On the other hand, most of the *very large dunes* ($\lambda_{bf} > 100$ m) were well aligned with both Chen et al. [13] (Equation (11)) and Flemming's (Equation (9)) lines and close to the equilibrium line, but several bedforms from FS–CNS2 and also a number from FS–CNS1 had a low wave height, corresponding to a steepness in the order of 0.01–0.02, so they may represent bedforms in adaptation. It is worth noting that during FS-CNS1, the longitudinal transect was collected after seven days of near-constant low-flow discharges, while FS-CNS2 was conducted during a period of continuously rising flow discharges. This could explain why the dune field may have mostly obtained a stable equilibrium with the flow conditions during FS-CNS1, while dune field was in a transitional phase during FS-CNS2 as it adjusted to the increasing flow discharge.

The length of the bedforms observed in the three field surveys were plotted against their steepness H_{bf}/λ_{bf} in Figure 13 and compared with the relationship proposed by Carling et al. [20]:

$$\frac{H_{bf}}{\lambda_{bf}} = 0.1027 \, \lambda_{bf}^{-0.615} \, (-) \tag{14}$$

and a new regression equation was derived:

$$\frac{H_{bf}}{\lambda_{bf}} = 0.208 \, \lambda_{bf}^{-0.437} \, (-) \tag{15}$$

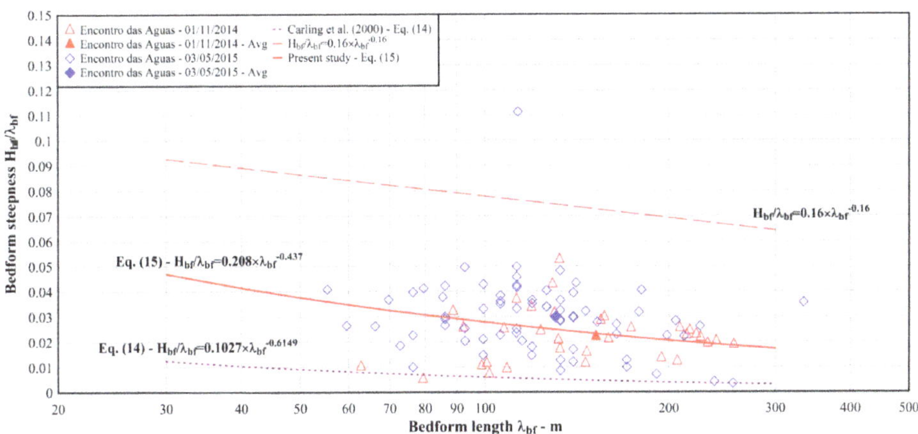

Figure 13. Bedform wavelength vs. bedform steepness.

Several bedforms, in both flow conditions, were characterised from steepness in the order of 0.01–0.2. This indicates that the bedforms with low steepness observed in FS–CNS2 were developing with the increasing discharge, while, on the other side, those in FS–CNS1 were in the process of crest flattening and elongation, having been formed during the previous high flow conditions [13].

As already mentioned above, dune sizes are often thought to scale with flow depth [10,11,17,26,51,52,55], as developing dunes cannot emerge out of the water [26], but, in the literature, other scaling relationships based on depth and grain size [24], transport stage [23,56] and transport stage and Froude number [57] were proposed. Transport stage is generally defined as any metric that is composed of a ratio of the shear stress to a grain size [25,26], including the Shields number and the Rouse number, which is defined as:

$$Ro = \frac{w_s}{\kappa\, u^*} \tag{16}$$

where w_s is particle settling velocity and u^* is the shear velocity, which can be obtained from the total bed shear stress τ_b as $u^* = (\tau_b/\rho)^{0.5}$. The most widely applied scaling equation for dune size is that of Yalin [51], where bedform wavelength and wave height are related to the flow depth as:

$$\lambda_{bf} = 5\,h\ (\text{m}) \tag{17}$$

$$H_{bf} = \frac{h}{6}\ (\text{m}) \tag{18}$$

while Yalin [52] suggested a theoretical value of $\lambda_{bf} = 5\,h$ for equilibrium dunes in deep flows. Venditti [55] analyzed Allen's [58] dataset and identified a range of variability between h and $16\,h$ for bedform length and between $1/40\,h$ and $1/6\,h$ for the bedform height. He also argued that the reason for this variability is that bedform sizes are dependent on transport stage and lag changes in the flow conditions.

Bradley and Venditti [26] re-evaluated seven predictive equations, including those from Yalin, linking dune dimensions to other variables such as flow depth, grain size, transport stage and Froude number. The data compilation using 498 observations coming from 21 flume experiments and 20 field studies shows that dune height and length follow a power law:

$$H_{bf} = 0.051\ \lambda_{bf}^{0.77}\ (\text{m}) \tag{19}$$

which is very similar to Flemming's equation (Equation (9)). Most of the data for bedform length and height were ranging from h to $16\,h$ and from $h/20$ to $h/2.5$, respectively. Bradley and Venditti [26] found that the predictive power of all the scaling relations was generally poor probably because these relations are not able to capture any effect of the flow variability over the time. They also observed that dunes in smaller channels conform to a different height scaling than dunes in larger channels, which reflects a change in dune morphology from strongly asymmetric dunes with high lee angles in flows <2.5 m deep to more symmetric, lower lee angle dunes in flows >2.5 m deep [26]. This implies a different process control rather than a continuum of processes as depth increases [26]. Hence, from the analysis of only data in channels with depth $h > 2.5$ m, they derived two different equations for wave height, a linear regression equation and a non-parametric scaling equation:

$$H_{bf} = 0.13\ h^{0.94}\ (\text{m}) \tag{20}$$

$$H_{bf} = \frac{h}{7.7}\ (\text{m}) \tag{21}$$

while, for all the data, they derived for wavelength again a linear regression equation and a non-parametric scaling equation:

$$\lambda_{bf} = 5.22\ h^{0.95}\ (\text{m}) \tag{22}$$

$$\lambda_{bf} = 5.9\ h\ (\text{m}) \tag{23}$$

Bradley and Venditti [26] recommended to apply non-parametric scaling relations, i.e., Equations (21) and (23), concluding that if it is clear that dunes increase its sizes with the scale of the river system, a sound explanation for how flow depth rules the equilibrium sizes of dunes is still lacking. Rather, the apparent scaling of dunes with flow depth may be only indirect and emerge because shear stress/velocity, both depending on depth, play a key role in dune morphology [26].

Figures 14 and 15 present the relationship between water depth and bedform wavelength and wave height, respectively, for the dunes observed in this study. The data were not exceeding the upper limits of scaling reported by [55]. On the other hand, while length data were quite well aligned between the curves for $\lambda_{bf} = 5.9\ h$ (Equation (23)) and $\lambda_{bf} = h$ (Figure 14), height data ranged mostly from $H_{bf} = h/6$ to $H_{bf} = h/40$, with some values above and below these curves (Figure 15). The bedforms with low wave height, i.e., $H_{bf} < h/20$ or $H_{bf} < h/40$, observed mostly in FS–CNS2 were probably adapting to the raising stage to high flow conditions. Finally, most of the wave height data were below the curve for $h/10$, confirming the findings observed in large rivers where height is often only 10% of the local flow depth [15].

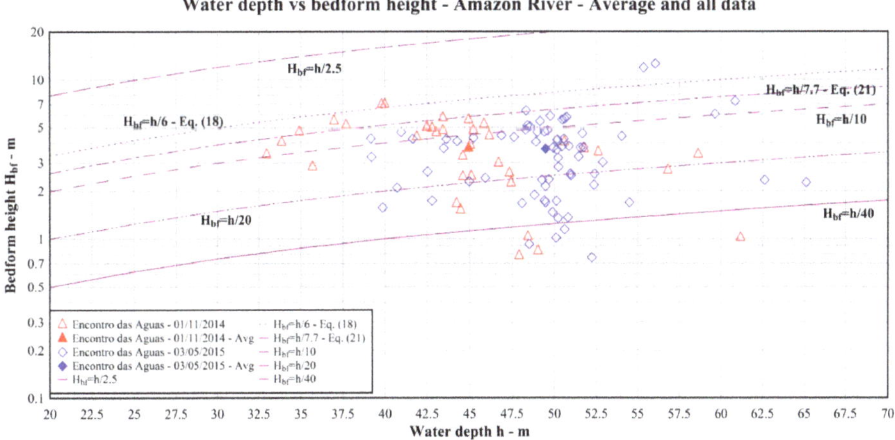

Figure 14. Bedform wavelength vs. water depth.

Figure 15. Bedform wave heigth vs. water depth.

Bedform sizes were also related to the transport stage expressed by the suspension number $u*/w_s$ [25], which is the inverse of the Rouse number (Equation (15)) multiplied by the von Kármán constant. Figure 16 presents the distribution of the relative bedform wavelength λ_{bf}/h and wave height H_{bf}/h against the suspension number $u*/w_s$, being in both flow conditions, in the range from 0.3 to 0.6. The lee side angle was below 10°, with an average value of 3.02° and 3.25° in 2014 and 2015 surveys, respectively. Lee side angles showed an interesting relationship to wave steepness (Figure 17). Bedform steepness grew gently with lee side angle and became constant above 6° at $H_{bf}/\lambda_{bf} = 0.05$, suggesting an interrelation between these parameters in this case [53]. Finally, the analysis of vertical velocity measured in the ADCP transects showed no flow separation due to the low lee side angles, as reported in past studies on low-angle dunes [1,15,59].

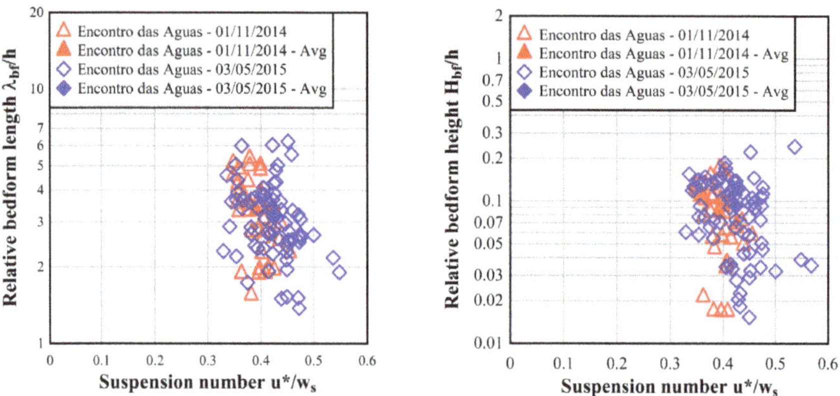

Figure 16. Suspension number $u*/w_s$ vs. λ_{bf}/h (left) and vs. H_{bf}/h (right).

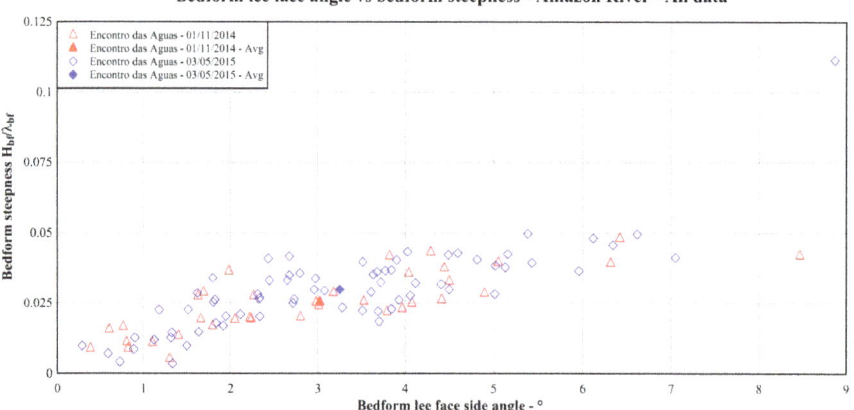

Figure 17. Bedform lee face angle vs. bedform steepness.

3.4. Comparison with Literature Data Sets from Large Rivers. Discussion

The field data collected is this study were compared with those from some literature data sets collected in large rivers. These data sets are: the side-scan sonar and sub-bottom profiler data collected in August 2003 in the middle–lower Changjiang (Yangtze) River (China) [13], and the multibeam echosounder data collected in January 2004 in the Lower Rhine (The Netherlands) [54], in May 2004 in the Rio Paraná (Argentina) [12], and in April 2008 in the Empire Reach of the lowermost Mississippi River (USA) [60]. Table 10 lists the main size parameters for these data sets.

Table 10. Main parameters of bedforms from the literature data sets. Legend: $\lambda_{bf\text{-}mean}$ = mean of the wavelength; $H_{bf\text{-}mean}$ = mean of the wave height; $(H_{bf}/\lambda_{bf})_{mean}$ = mean of the wave steepness.

Data Set	n.	$\lambda_{bf\text{-}mean}$ (m)	$H_{bf\text{-}mean}$ (m)	$(H_{bf}/\lambda_{bf})_{mean}$ (%)
Changjiang River—August 2003	138	79.19	2.35	5.95
Lower Rhine—January 2004	61	11.89	0.58	5.31
Rio Paraná—May 2004	36	53.92	1.50	3.05
Mississippi River—April 2008	12	62.33	1.69	3.70

Figure 18 presents the relationship between bedform wavelength and bedform wave height for the dunes observed in the present study and those reported in the above literature data sets along with the empirical regression laws in Equations (9)–(13). In the range of *medium dunes* ($5.0 > \lambda_{bf} > 10$ m) or more, the data from the Lower Rhine and Mississippi River were aligned with Equation (9), while the data from the Changjiang River were proximal to Flemming's maximum line (Equation (10)) and their steepness was larger than 0.06. Moreover, several data from the Lower Rhine were below Flemming's line (Equation (9)). In the range of *large dunes* ($10.0 > \lambda_{bf} > 100$ m), most of the dunes from Rio Paraná and several from the Changjiang River were aligned with Equation (13) and close to the equilibrium line, but a number of dunes from Rio Paraná were below Flemming's line (Equation (9)). In the range of *very large dunes* ($\lambda_{bf} > 100$ m), most of the literature data were close to the Amazon River data and it is possible to observe that a number of data from the Changjiang River were characterised from low wave height and steepness in the order of 0.01, as they were probably adapting to changes in flow conditions.

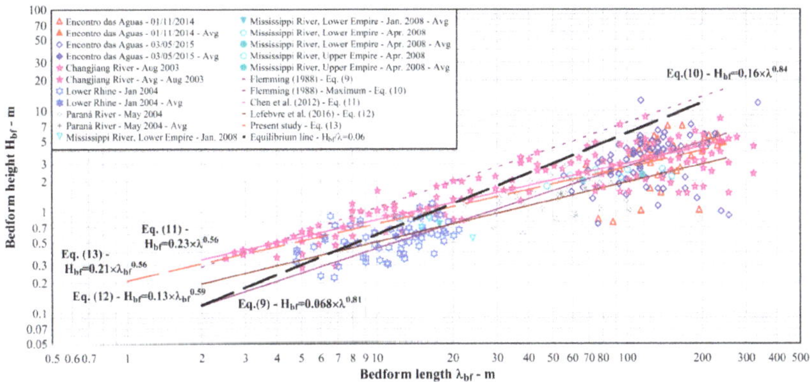

Figure 18. Bedform wavelength vs. bedform wave height, comparison with literature data from large rivers.

4. Conclusions

In alluvial rivers, the geometry of the bed topography is the result of a complex interaction among several hydrodynamics and sedimentary processes acting under the constraint of varying boundary conditions. In sand-bedded alluvial channels, the bottom boundary consists of a labile bed comprising bedforms of many different scales and geometries. Bedforms are deformations of a sand bed that are smaller than channel-scale bar forms and that have specific geometric properties [26,55]. They also are subjected to deformation and adaptation to changes in river flow [9]. This paper presented the results of a study about the morphology of the bedforms observed in the area of the Negro/Solimões confluence. Two surveys using acoustic Doppler velocity profiling (ADCP) were carried out in low flow

(FS-CNS1) conditions, after seven days of near-constant discharge, and in relatively high flow (FS-CN2) conditions, during a period of continuously rising flow discharges. The observed bedforms were mostly in the range of *large* and *very large dunes* according to Ashley [22] classification with a maximum wavelength and wave height of 350 and 12 m, respectively. Second, during (FS-CN2), maximum bedform sizes as well as in the frequency distribution of bedform size were comparatively larger, as could be expected from the literature studies. Third, most of the *large dunes* (10.0 > λ_{bf} > 100 m) and *very large dunes* (λ_{bf} > 100 m) were generally in equilibrium with flow conditions. On the other side, some bedforms observed in relatively high flow conditions were developing to adjust to the continuously increasing flow discharge, while only some in low flow conditions were in the process of crest flattening and elongation, having been formed during the previous high flow conditions. Fourth, the data were within the upper limits of scaling with water depth reported in the literature, but, while length data were quite well aligned with scaling curves, height data showed a scatter from these curves. Fifth, in both surveys, the lee side angle was below 10° with an average value of about 3.0° and no flow separation was observed confirming recent literature studies on large rivers [15], and wave steepness grew gently with lee side angle and became constant above 6° at $H_{bf}/\lambda_{bf} = 0.05$. Finally, a comparison between the data collected in this study and past literature studies on large rivers suggested that a number of the largest bedforms were probably adapting to discharge variations in the river.

Author Contributions: All the authors contributed to this study. Analysis and investigation, C.G., I.M., N.P.F.J. and M.I.; data processing, C.G. and I.M.; writing—original draft preparation, C.G.; writing—review and editing, I.M., N.P.F.J. and M.I.; visualization, C.G., I.M. and M.I.; funding acquisition, C.G., N.P.F.J. and M.I. All authors have read and agreed to the published version of the manuscript.

Funding: This research was carried out within the Clim-Amazon Research Project funded by grant agreement FP7 INCO-LAB n°295091 from the European Commission. Naziano Filizola acknowledges funding by CAPES-Procad Amazonia and CNPq-AmazonGeoSed Project.

Acknowledgments: The authors acknowledge Mark Trevethan for his valuable collaboration in the Clim-Amazon Project and the CPRM (Geological Survey of Brasil) for supplying the research vessel, instrumentation, technical assistance with sampling; Bosco Alfenas, Andrè Martinelli Santos, Daniel Moreira, Arthur Pinheiro, Paulo Melo, Nilda Pantoja, Andre Zumak and João Andrade for their assistance with sampling during the surveys; and finally Farhad Bahmanpouri for his assistance in the extraction of the ADCP data.

Conflicts of Interest: The authors declare no conflict of interest.

References

1. Best, J.L. The fluid dynamics of river dunes: A review and some future research directions. *J. Geophys.* **2005**, *110*, F04S02. [CrossRef]

2. Chanson, H. *The Hydraulics of Open Channel Flow: An Introduction*, 2nd ed.; Butterworth-Heinemann: Oxford, UK, 2004; p. 630.

3. Andreotti, B.; Claudin, P.; Devauchelle, O.; Duràn, O.; Fourrière, A. Bedforms in a turbulent stream: Ripples, chevrons and antidunes. *J. Fluid Mech.* **2012**, *690*, 94–128. [CrossRef]

4. Stoesser, T.; Braun, C.; Garcìa-Villalba, M.; Rodi, W. Turbulence Structures in Flow over Two-Dimensional Dunes. *J. Hydraul. Eng.* **2008**, *134*, 42–55. [CrossRef]

5. Omidyeganeh, M.; Piomelli, U. Large-eddy simulation of three-dimensional dunes in a steady, unidirectional flow. Part 1. Turbulence statistics. *J. Fluid Mech.* **2013**, *721*, 454–483. [CrossRef]

6. Omidyeganeh, M.; Piomelli, U. Large-eddy simulation of three-dimensional dunes in a steady, unidirectional flow. Part 2. Flow structures. *J. Fluid Mech.* **2013**, *734*, 509–534. [CrossRef]

7. Bennett, S.J.; Best, J.L. Mean flow and turbulence structure over fixed, two dimensional dunes: Implications for sediment transport and bedform stability. *Sedimentology* **1995**, *42*, 491–513. [CrossRef]

8. Ojha Satya, P.; Mazumder, B.S. Turbulence characteristics of flow over a series of 2-D bed forms in the presence of surface waves. *J. Geophys. Res. Earth Surf.* **2010**, *115*, F404016.

9. Reesink, A.J.H.; Parsons, D.R.; Ashworth, P.J.; Best, J.L.; Hardy, R.J.; Murphy, B.J.; McLelland, S.J.; Unsworth, C. The adaptation of dunes to changes in river flow. *Earth-Sci. Rev.* **2018**, *185*, 1065–1087. [CrossRef]

10. Bradley, R.W.; Venditti, J.G. Transport scaling of dune dimensions in shallow flows. *J. Geophys. Res. Earth Surf.* **2019**, *124*, 526–547. [CrossRef]

11. Bradley, R.W.; Venditti, J.G. The growth of dunes in rivers. *J. Geophys. Res. Earth Surf.* **2019**, *124*, 548–566. [CrossRef]

12. Parsons, D.R.; Best, J.L.; Orfeo, O.; Hardy, R.J.; Kostaschuk, R.; Lane, S.N. Morphology and flow fields of three-dimensional dunes, Rio Paranà, Argentina: Results from simultaneous multibeam echo sounding and acoustic Doppler current profiling. *J. Geophys. Res. Earth Surf.* **2005**, *110*, F04S03. [CrossRef]

13. Chen, J.; Wang, Z.; Li, M.; Wei, T.; Chen, Z. Bedform characteristics during falling flood stage and morphodynamic interpretation of the middle-lower Changjiang (Yangtze) River channel, China. *Geomorphology* **2012**, *147–148*, 18–26. [CrossRef]

14. Bialik, R.J.; Karpiński, M.; Rajwa, A.; Luks, B.; Rowiński, P.M. Bedform Characteristics in Natural and Regulated Channels: A Comparative Field Study on the Wilga River, Poland. *Acta Geophys.* **2014**, *62*, 1413–1434. [CrossRef]

15. Cisneros, J.; Best, J.L.; van Dijk, T.; Paes de Almeida, R.; Amsler, M.; Boldt, J.; Freitas, B.; Galeazzi, C.; Huizinga, R.; Ianniruberto, M.; et al. Dunes in the world's big rivers are characterized by low-angle lee-side slopes and a complex shape. *Nat. Geosci.* **2020**, *13*, 156–162.

16. Julien, P.Y.; Klaassen, G.J.; Ten Brinke, W.B.M.; Wilbers, A.W.E. Case Study: Bed Resistance of Rhine River during 1998 Flood. *J. Hydraul. Eng.* **2002**, *128*, 1042–1050. [CrossRef]

17. Cheel, R.J. *Introduction to Clastic Sedimentology, Course Notes*; Brock University: St. Catharines, ON, Canada, 2005.

18. Almeida, R.P.; Galeazzi, C.P.; Freitas, B.T.; Janikian, L.; Ianniruberto, M.; Marconato, A. Large barchanoid dunes in the Amazon River and the rock record: Implications for interpreting large river systems. *Earth Planet. Sci. Lett.* **2016**, *454*, 92–102. [CrossRef]

19. Simons, D.B.; Richardson, E.V. *Resistance to Flow in Alluvial Channels*; Technical Report for Geological Survey Professional Paper 422-J; United States Government Printing Office: Washington, DC, USA, 1961.

20. Carling, P.A.; Gölz, E.; Orr, H.G.; Radecki-Pawlik, A. The morphodynamics of fluvial sand dunes in the River Rhine, near Mainz, Germany. I. Sedimentology and morphology. *Sedimentology* **2000**, *47*, 227–252. [CrossRef]

21. Yalin, M.S. *The Mechanics of Sediment Transport*, 2nd ed.; Pergamon Press: Oxford, UK, 1977; p. 298.

22. Ashley, G.M. Classification of large-scale subaqueous bedforms: A new look at an old problem. *J. Sediment Res.* **1990**, *60*, 160–172.

23. Van Rijn, L.C. Sediment Transport, Part III: Bed forms and Alluvial Roughness. *J. Hydraul. Eng.* **1984**, *110*, 1733–1754. [CrossRef]

24. Julien, P.Y.; Klaassen, G.J. Sand-dune geometry of large rivers during floods. *J. Hydraul. Eng.* **1995**, *121*, 657–663. [CrossRef]

25. Naqshband, S.; Ribberink, J.S.; Hulscher, S.J. Using both free surface effect and sediment transport mode parameters in defining the morphology of river dunes and their evolution to upper stage plane beds. *J. Hydraul. Eng.* **2014**, *140*, 06014010. [CrossRef]

26. Bradley, R.W.; Venditti, J.G. Reevaluating dune scaling relations. *Earth-Sci. Rev.* **2017**, *165*, 356–376. [CrossRef]

27. Kostaschuk, R.; Best, J. Response of sand dunes to variations in tidal flow: Fraser Estuary, Canada. *J. Geophys. Res. Earth Surf.* **2005**, *110*, F04S04. [CrossRef]

28. Maddux, T.B.; Nelson, J.M.; McLean, S.R. Turbulent flow over three-dimensional dunes: 1. Free surface and flow response. *J. Geophys. Res. Earth Surf.* **2003**, *108*, 6009. [CrossRef]

29. Maddux, T.B.; Nelson, J.M.; McLean, S.R. Turbulent flow over three-dimensional dunes: 2. Free surface and flow response. *J. Geophys. Res. Earth Surf.* **2003**, *108*, 6010. [CrossRef]

30. Best, J.L.; Rhoads, B.L. Sediment transport, bed morphology and the sedimentology of river channel confluences. In *River Confluences, Tributaries and the Fluvial Network*; Rice, S., Roy, A., Rhoads, B., Eds.; John Wiley & Sons, Ltd.: New York, NY, USA, 2008; pp. 45–72.

31. Kentworthy, S.; Rhoads, B.L. Hydrologic control of spatial patterns of suspended sediment concentration at a stream confluence. *J. Hydrol.* **1995**, *168*, 251–263. [CrossRef]

32. Best, J. Flow dynamics at river channel confluences: Implications for sediment transport and bed morphology. In *Recent Developments in Fluvial Sedimentology*; Ethridge, F., Flores, M., Harvey, M., Eds.; Society of Economic Paleontologists and Mineralogists: Tulsa, OK, USA, 1987; Volume 39, pp. 27–35.

33. Biron, P.; Lane, S. Modelling hydraulics and sediment transport at river confluences. In *River Confluences, Tributaries and the Fluvial Network*; Rice, S., Roy, A., Rhoads, B., Eds.; John Wiley & Sons, Ltd.: New York, NY, USA, 2008; pp. 17–43.

34. Filizola, N.; Spinola, N.; Arruda, W.; Seyler, F.; Calmant, S.; Silva, J. The Rio Negro and Rio Solimões confluence point—Hydrometric observations during the 2006/2007 cycle. In *River, Coastal and Estuarine Morphodynamics: RCEM 2009*, 1st ed.; Vionnet, C., Garcia, M.H., Latrubesse, E.M., Perillo, G.M.E., Eds.; Taylor & Francis Group: London, UK, 2009.

35. Guyot, J.L. *Hydrogéochimie des Fluves de L'Amazonie Bolivienne*; ORSTOM: Paris, France, 1993.

36. Laraque, A.; Guyot, J.; Filizola, N. Mixing processes in the Amazon River at the confluences of the Negro and Solimões Rivers, Encontro das Aguas, Brazil. *Hydrol. Process.* **2009**, *23*, 3131–3140. [CrossRef]

37. Filizola, N.; Guyot, J.L. Suspended sediment yields in the Amazon basin: An assessment using the Brazilian national data set. *Hydrol. Process.* **2009**, *23*, 3207–3215. [CrossRef]

38. Trevethan, M.; Martinelli, A.; Oliveira, M.; Ianniruberto, M.; Gualtieri, C. Fluid dynamics, sediment transport and mixing about the confluence of Negro and Solimões rivers, Manaus, Brazil. In Proceedings of the 36th IAHR World Congress, The Hague, The Netherlands, 28 June–3 July 2015; p. n80094.

39. Trevethan, M.; Ventura Santos, R.; Ianniruberto, M.; Santos, A.; De Oliveira, M.; Gualtieri, C. Influence of tributary water chemistry on hydrodynamics and fish biogeography about the confluence of Negro and Solimões rivers, Brazil. In Proceedings of the 11th International Symposium on EcoHydraulics (ISE 2016), Melbourne, Australia, 2–7 February 2016; p. 25674.

40. Gualtieri, C.; Ianniruberto, M.; Filizola, N.; Ventura Santos, R.; Endreny, T.A. Hydraulic complexity at a large river confluence in the Amazon Basin. *EcoHydrology* **2017**, *10*. [CrossRef]

41. Gualtieri, C.; Filizola, N.; Oliveira, M.; Santos, A.M.; Ianniruberto, M. A field study of the confluence between Negro and Solimões Rivers. Part 1: Hydrodynamics and sediment transport. *C. R. Geosci.* **2018**, *350*, 31–42. [CrossRef]

42. Gualtieri, C.; Ianniruberto, M.; Filizola, N. On the mixing of rivers with a difference in density: The case of the Negro/Solimões confluence, Brazil. *J. Hydrol.* **2019**, *578*. [CrossRef]

43. Ianniruberto, M.; Trevethan, M.; Pinheiro, A.; Andrade, J.F.; Dantas, E.; Filizola, N.; Santos, A.; Gualtieri, C. A field study of the confluence between Negro and Solimões Rivers. Part 2: River bed morphology and stratigraphy. *C. R. Geosci.* **2018**, *350*, 43–54. [CrossRef]

44. Szupiany, R.N.; Amsler, M.L.; Parsons, D.R.; Best, J.L. Morphology, flow structure and suspended bed sediment transport at large braid-bar confluences. *Water Resour. Res.* **2009**, *45*, W05415. [CrossRef]

45. Kostaschuk, R.; Villard, P.; Best, J. Measuring velocity and shear stress over dunes with Acoustic Doppler Profiler. *J. Hydraul. Eng.* **2004**, *130*, 932–936. [CrossRef]

46. Sime, L.C.; Ferguson, R.I.; Church, M. Estimating shear stress from moving boat acoustic Doppler velocity measurements in a large gravel bed river. *Water Resour. Res.* **2007**, *43*, W03418. [CrossRef]

47. Wilcock, P.R. Estimating local shear stress from velocity observations. *Water Resour. Res.* **1996**, *32*, 3361–3366. [CrossRef]

48. Nordin, C.F.; Meade, R.H., Jr.; Mahoney, H.A.; Delaney, B.M. *Particle Size of Sediments Collected from the Bed of the Amazon River and Its Tributaries in June and July 1976*; United States Department of the Interior Geological Survey: Denver, CO, USA, 1977.

49. Trevethan, M.; Aoki, S. Initial Observations on Relationship between Turbulence and Suspended Sediment Properties in Hamana Lake Japan. *J. Coast. Res.* **2009**, *56*, 1434–1438.

50. Bahmanpouri, F.; Filizola, N.; Ianniruberto, M.; Gualtieri, C. A new methodology for presenting hydrodynamics data from a large river confluence. In Proceedings of the 37th IAHR World Congress, Kuala Lumpur, Malaysia, 13–18 August 2017.

51. Yalin, M.S. Geometrical properties of sand waves. *J. Hydraul. Div.* **1964**, *90*, 105–119.

52. Yalin, M.S. *River Mechanics*, 1st ed.; Pergamon Press: New York, NY, USA, 1992.

53. Lefebvre, A.; Paarlberg, A.J.; Winter, C. Characterising natural bedform morphology and its influence on flow. *Geo-Mar. Lett.* **2016**, *36*, 379–393. [CrossRef]

54. Frings, R.M. From Gravel to Sand. Downstream Fining of Bed Sediments in the Lower River Rhine. Ph.D. Thesis, Utrecht University, The Netherlands Geographical Studies 368, Royal Dutch Geographical Society, Utrecht, The Netherlands, 2007.

55. Venditti, J.G. Bedforms in sand-bedded rivers. In *Treatise on Geomorphology*; Shroder, J., Wohl, E., Eds.; Academic Press: San Diego, CA, USA, 2013; Volume 9, pp. 137–162.

56. Karim, F. Bed configuration and hydraulic resistance in alluvial channel flows. *J. Hydraul. Eng.* **1995**, *121*, 15–25. [CrossRef]

57. Karim, F. Bed-form geometry in sand-bed flows. *J. Hydraul. Eng.* **1999**, *125*, 1253–1261. [CrossRef]
58. Allen, J.R.L. *Sedimentary Structures: Their Character and Physical Basis*, 1st ed.; Elsevier: Amsterdam, The Netherlands, 1982.
59. Lefebvre, A.; Paarlberg, A.J.; Ernstsen, V.B.; Winter, C. Flow separation and roughness lengths over large bedforms in a tidal environment: A numerical investigation. *Cont. Shelf Res.* **2014**, *91*, 57–69. [CrossRef]
60. Nittrouer, J.A. Sediment transport dynamics in the Lower Mississippi River: Non-uniform Flow and Its Effects on Tiver Channel Morphology. Ph.D. Thesis, University of Texas at Austin, Austin, TX, USA, 2010.

Article

Man-Induced Discrete Freshwater Discharge and Changes in Flow Structure and Bottom Turbulence in Altered Yeongsan Estuary, Korea

KiRyong Kang [1] and Guan-hong Lee [2,*]

[1] National Institute of Meteorological Sciences/KMA, Seogwipo 63568, Korea; krkang@korea.kr
[2] Department of Oceanography, Inha University, Incheon 22212, Korea
* Correspondence: ghlee@inha.ac.kr

Received: 3 June 2020; Accepted: 1 July 2020; Published: 5 July 2020

Abstract: Flow measurements were performed in the altered Yeongsan estuary, Korea, in August 2011, to investigate changes in flow structure in the water column and turbulence characteristics very close to the bed. Comparison between the bottom turbulent kinetic energy (TKE) and suspended sediment concentration (SSC) was conducted to examine how discrete freshwater discharge affects the bottom sediment concentration. The discrete freshwater discharge due to the gate opening of the Yeongsan estuarine dam induced a strong two-layer circulation: an offshore-flowing surface layer and a landward-flowing bottom layer. The fine flow structure from the bed to 0.35 m above the bottom (mab hereafter) exhibited an upside-down-bell-shaped profile for which current speed was nearly uniform above 0.1 mab, with the magnitude of the horizontal and vertical flow speeds reaching 0.1 and 0.01 m/s, respectively. The bottom turbulence responded to the freshwater discharge at the surface layer and the maximum magnitude of the Reynolds stress reached up to 2×10^{-4} m^2/s^2 during the discharged period, which coincided with increased SSC in the bottom boundary layer. These results indicate that the surface freshwater discharge due to opening of the estuarine dam gate increases the SSC by the discharge-induced intensification of the turbulent flow in the bottom boundary layer.

Keywords: Yeongsan estuary; freshwater discharge; two-layer circulation; Reynolds stress; bottom turbulence; suspended sediment concentration

---

1. Introduction

In estuaries, water currents are mainly driven by tide, wind, and freshwater discharge [1]. Other factors affecting the flow structure include density stratification caused by temperature and salinity, seabed roughness due to the bottom surficial texture and bedforms, and also artificial alterations such as dam construction. Water currents play an important role in the transportation and distribution of suspended sediment which can change the topography and morphological shape of an estuary. Especially when the bottom surface is composed of silt or mud, resuspension of sediment can be very dependent on the turbulent flow activity above the bed. Turbulent flow near the bed is a well-known factor which generates sediment-related processes such as erosion, dispersion, and transportation in the bottom layer [2–4].

Artificial alteration of an estuarine environment, such as dam construction to block the saltwater intrusion and the regulation of the freshwater discharge, can modify the physical environment such as the tidal range and circulation structure [5,6]. One factor which can induce a sudden response from an estuary is forced freshwater discharge because rapid currents can be formed in the surface with a velocity difference between the surface and bottom layers. When any such artificial change occurs such as discontinuous freshwater input, an estuary naturally will adjust to these changes. An example of an estuary which has experienced such artificial change is the Yeongsan River estuary (Figure 1), located in the southwestern portion of the Korean peninsula. Yeongsan River is one of the four major rivers in Korea. It has a drainage area of 3455 km^2 and a length of 129.5 km. The Yeongsan estuary has a series of offshore islands and extensive salt marsh complexes, with an average water depth of 20 m. The main channel was blocked by a dam in February 1981, and now it behaves as a semi-enclosed bay [7]. The tide is macrotidal with a semi-diurnal tidal range of 4.5 m, and 70% of the annual rainfall of 840 mm occurs from June to August.

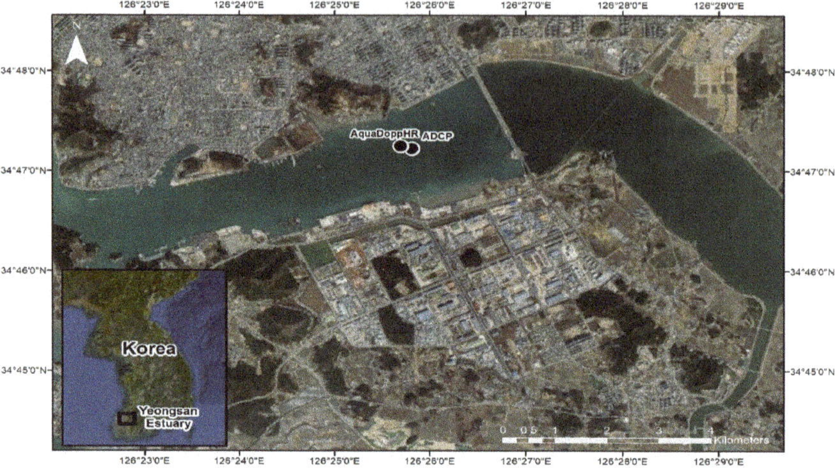

Figure 1. Study area and location of current meters in Yeongsan estuary, Korea. Black dots mark the location of two current meters placed in August 2011.

According to several studies, the physical environmental conditions have changed after the dam construction. Kang [6] showed that the tidal range increased to 60 and 43 cm for extreme high and low tide, respectively, while the tidal velocity decreased with an ebb tidal dominance causing a change in the sediment transport mechanisms. During the summer season when the gate opens frequently, the freshwater discharge has become an important factor to change the current, temperature, and salinity distribution because of sudden and forced discharge through the surface layer [8–12]. In terms of flow system change, for example, Cho et al. [11] suggested that there exist four layers under low discharge conditions during the summer season, showing seaward flow in the surface and middle layer, and landward flow in the bottom layer and between the surface and middle layers. With freshwater discharge due to the gate opening, a two-layer circulation is formed with strong stratification between the offshore-flowing surface layer and the landward-flowing bottom layer. So, the freshwater discharge can affect directly the estuarine circulation system. In addition, through ^{210}Pb and ^{7}Be radioisotope geochronology in the Yeongsan estuary, Williams et al. [13] showed that high sedimentation rates up to 9 cm/year occur in the estuary, and the sediment deposition primarily occurs during episodic events corresponding to high discharge. However, the flow structure and the associated sediment suspension near the bed during freshwater discharge are still not well understood.

In the Yeongsan estuary, a sudden release of freshwater in the surface layer due to opening of the estuarine dam gate can cause rapid intensification of the seaward flow. It is expected, then, that the flow of the lower layer should respond to the sudden and strong surface flow. In such a case, the bottom flow response could be coupled to change in the velocity shear, and this in turn could be a cause for sediment resuspension from the bed. To our knowledge, this is the first field observation to report the characteristics of flow, turbulence, and suspended sediment in the bottom boundary layer in response to discrete freshwater discharge. Somewhat related to this topic includes those of entrainment and mixing in a turbulent jet [14,15], jet scour [16], and sediment discharge by jet-induced flow [17–19].

Therefore, the main objectives in this study are to investigate how the freshwater discharge changes the structure of the bottom turbulent flow properties like the Reynolds stress and the turbulent kinetic energy (TKE), and to elucidate how the bottom turbulent flow interacts with the suspended sediment near the seabed by using observed water current and suspended sediment concentration data to capture the turbulent structure and suspended sediment concentration in the bottom boundary layer during a freshwater discharge. The observation and data processing scheme are described in Section 2, and the analysis results about the evolution of the mean and turbulent flow and the bottom suspended sediment concentrations are shown in Section 3. The role of man-induced freshwater discharge on the flow structure and suspended sediment concentration fluctuations above the seabed is discussed in Section 4, and finally, a very brief conclusion and meaning of this study are shown in Section 5.

2. Observation and Data Processing

The experiment campaign was designed specifically to observe the flows in both the bottom and upper layers during the period of discrete freshwater discharge in the Yeongsan estuary, Korea, during August 2011 (Figure 1). Two current meters, an ADCP (RDI 1200 kHz, Teledyne, Poway, CA) and an AquaDoppHR (Model: AQP 9116, Nortek, Boston, MA), were moored near the Yeongsan estuarine dam in the inner estuary (Figure 1 and Table 1). The ADCP was mounted on the bed with an up-looking orientation to measure the flow profile in the water column, and the AquaDoppHR was bottom-mounted at about 1 m above bottom (mab hereafter) with down-looking orientation for near-bed turbulence measurements. The current profiles obtained from the ADCP with high percent good values of 85 and above were averaged by burst with a 30-min interval for the flow structure in the water column (Table 1).

Table 1. Measurement scheme for the current meters in Yeongsan estuary.

Instrument	Location	Sampling Rate/Interval/ Period	Bin Size/ Blank Dist./ Orientation	Remarks
ADCP Sentinel (RDI 1200kHz)	34°47′13.47″ N 126°25′48.88″ E	2 Hz/30 min/ 5–28 August 2011	0.25 m/0.38 m/ Up-looking	Frame height: 0.7 m
AquaDoppHR (AQP 9116)	34°47′15.05″ N 126°25′41.85″ E	4 Hz/30 min/ 23–29 August 2011	0.05 m/0.40 m/ Down-looking	

The ADCP current data were not rotated into the along- and cross-channel directions because the main direction of the freshwater discharge was east–west. The observed current data were separated into tidal and residual flows based on the tidal harmonic analysis developed by Foreman [20] and implemented into MATLAB as T_TIDE [21]. From the residual flow, it was possible to check how the upper layer flow responded to the sudden freshwater discharge from the gate opening of the dam.

The AquaDoppHR was programmed for burst sampling to observe the fine structure of the mean and turbulent flow characteristics very near the bed. The burst sampling lasted for 8.5 min in 30-min burst intervals the same as the ADCP, however, the AquaDoppHR sampling frequency was higher at 4 Hz. The AquaDoppHR profiled from the bed to 0.35 mab with 5 cm bin size spatial resolution. The bed was detected from the trend of the acoustic signal strength along each beam, in which the signal strength generally shows maximum value at the bed. In this case, since three beams were looking at the bottom, the maximum value was located in different bins of each beam such as the 7th, 8th, and 9th bins. A possible reason for different bin numbers showing the maximum signal strength could be the bed status. Since the bed surface that the AquaDoppHR was looking down upon was not perfectly flat, and because it is hard to know the true bed status at the midpoint between the three beams, we selected the middle value of bins as the bed level.

The near-bed velocity data obtained from the AquaDoppHR were first despiked using an averaging method after visual inspection of all bursts. Then, the mean, variance, and covariance values of the different flow components of all bins for each burst were calculated to estimate the turbulent flow characteristics such as the Reynolds stress components and the turbulent kinetic energy (TKE). The Reynolds stress and the TKE were calculated by using the following two equations (Equations (1) and (2)):

$$R_{13} = -<u'w'>, \; R_{23} = -<v'w'> \tag{1}$$

$$TKE = (<u'^2> + <v'^2> + <w'^2>)/2 \tag{2}$$

where u', v', and w' are the turbulent components of the east–west (u), north–south (v), and up–down (w) components of a velocity vector, $u = (u, v, w)$, for each burst sampling period, respectively. U and V are the mean flow components of each burst for horizontal flow, for example, $u = U + u'$ and $v = V + v'$, and the bracket ($< >$) means time averaging for a burst period. Finally, the suspended sediment concentration (SSC) was calculated by conversion of the acoustic backscatter signal of the four beams of the moored ADCP. A more detailed procedure for this calculation can be found in Park and Lee [22].

3. Results

3.1. Freshwater Discharge and Flow Structure Change

One of the key factors to affect the estuarine flow in an altered estuary is the freshwater discharge since it makes both a forced seaward current in the surface layer and a density difference between the upper and lower layers. Moreover, the strength of the surface current as well as the density gradient depends on the amount of freshwater discharge. In this section, the response of the surface water after the freshwater discharge is depicted using the ADCP data. A time series of discharge is compared with the current structure and then with the change in residual flow (component with tides removed, using 17 tidal constituents) in the water column.

Figure 2 shows the discrete freshwater discharge and mean current velocity for U and V, during 5–28 August 2011. According to typical gate operation procedure, the gates were opened only during the low tide to prevent saltwater intrusion. On 11 August, over 6×10^7 tons of freshwater were discharged. After the discharge, the westward surface current rapidly increased to greater than 0.5 m/s and affected at least 5 m below the surface. At the same time, an opposite flow to the east formed in the lower layer. The sudden release of freshwater also increased the north–south component of the flow in a short period of time. When there was no freshwater discharge, the current speed was less than 0.25 m/s and the flow structure simply repeated the flood and ebb states such that eastward flow occurred during flood, westward flow occurred during ebb, and almost zero flow occurred during both high and low slack tides.

The tide-removed residual flow showed in detail how the mean flow structure could be changed by the amount of freshwater discharge (Figure 3). During freshwater discharge, the residual flow showed a two-layer system as expected wherein the upper layer moves seaward and the lower layer flows landward. The flow was seaward from the surface to 0.6 z/H (i.e., z = 9.6 m for H = 16 m) on 10 August 2011, and the depth of the seaward flow decreased as the freshwater discharge decreased. When the amount of discharged water increased, the lower layer also showed a stronger response of landward flow just after the discharge. Of course, the peak speed of the residual u-component appeared in the surface. During no or very weak discharge periods (20–22 August in Figure 3), the residual flow was also weak (almost less than 0.01 m/s) and the vertical structure was hard to specify because of the multilayered structure. This stagnant vertical structure changed with a rapid increase in the speed of the residual current when freshwater discharges occurred.

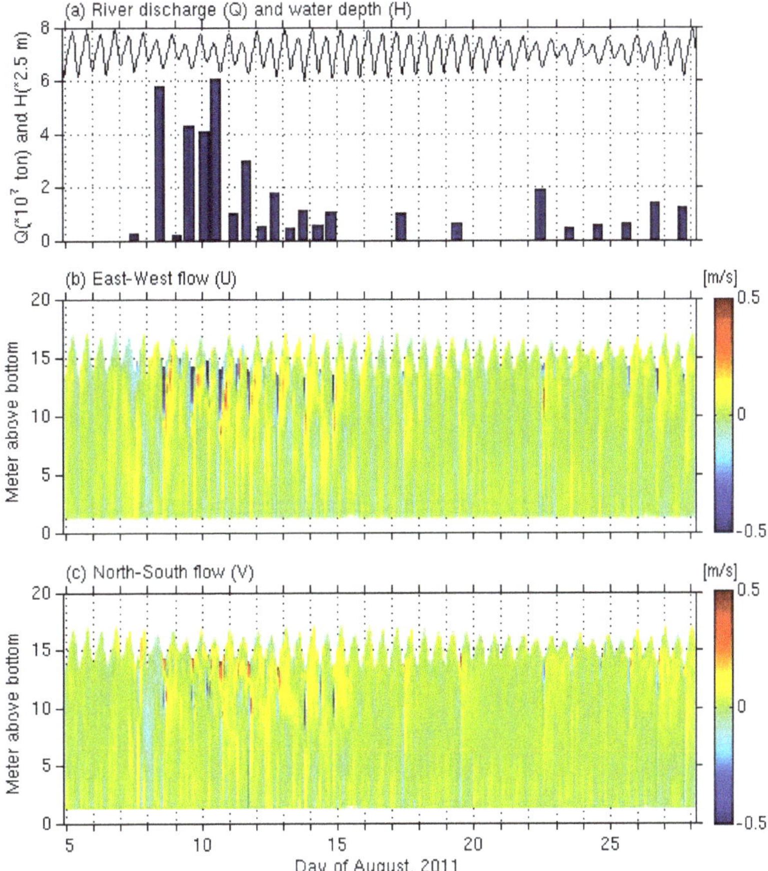

Figure 2. Time series of the freshwater discharge and water flow. (**a**) Freshwater discharge and water depth, (**b**) u-component of current velocity, (**c**) v-component of current velocity. In (**b**,**c**), negative values indicate west- and south-directed velocities, respectively.

The averaged residual flow structure for the whole observation period was a typical two-layer flow system that should be induced by vertical gravitational circulation in an estuary [1,23]. The mechanism of this structure is known very well, and there are two forcing factors: the barotropic and baroclinic forcing. The sea level difference between the upper and lower estuary acts as the barotropic forcing, making seaward flow in the upper layer, while the density difference forms the baroclinic forcing, making landward flow in the lower layer. The v-component showed a single-layer structure with very weak flow to the north, and the speed was less than 2 cm/s. This could have been due to the position of the dam gate which is located at the southern part of the dam (Figure 1). The two-layer structure during the entire period was an important aspect of the overall vertical structure of the flow since there was a stagnant, multilayer flow structure during the weak or no freshwater discharge periods (Figure 3d,e).

The strength of the upper- and lower-layer flow depended on the amount of the freshwater discharge. Figure 3 shows the vertical structure change of the residual flow with variation in the amount of discharge, for example, relatively high, low, and no freshwater cases. During the high freshwater discharge period on 8–11 August 2011, the maximum speed of the seaward residual flow was about 0.1 m/s in the surface layer, and a landward flow existed below 0.7 z/H with a quarter of the speed of the surface layer. As the amount of freshwater discharge decreased, the thickness of landward flow in the subsurface layer was increased from 0.7 z/H to 0.8 z/H, and the speed of the surface layer rapidly decreased. This represents the seawater response inside the estuary after blocking the freshwater release. When the barotropic forcing completely disappears due to no freshwater discharge, the water column has a multilayer flow structure, for example, a three- or four-layer structure, indicating that there were many local flows without a dominant flow pattern (Figure 3e). This indicated that during the no freshwater discharge period in this estuary, the residual flow pattern was indistinct and exhibited weak and vertically variable horizontal currents in a stagnant environment. Therefore, the freshwater discharge plays the major role in intensifying the two-layer system, including the residual circulation, in this estuary.

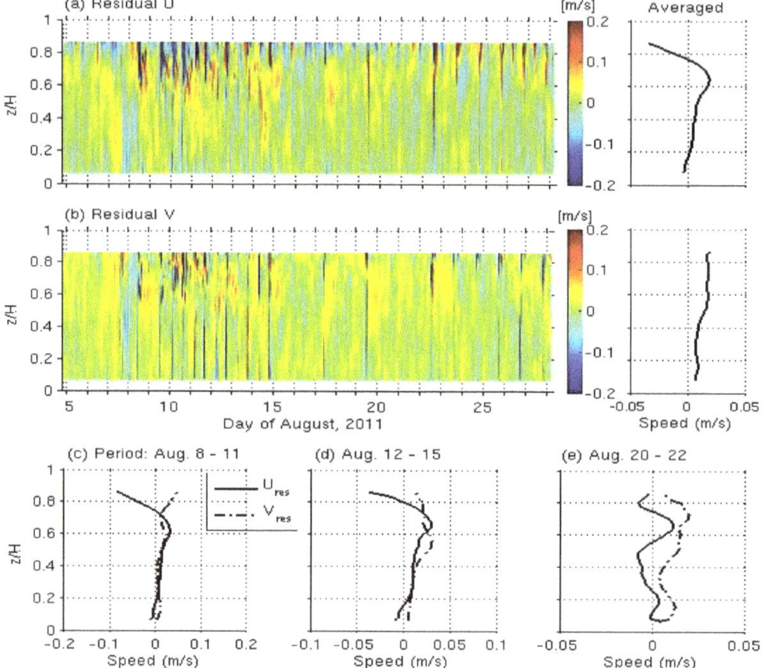

Figure 3. Change of vertical structure of the residual (tide-removed) flow: (**a**) east–west component, (**b**) north–south component, and mean flow structure of each period such as (**c**) high (i.e., $Q > 4 \times 10^7$ ton on 8–11 August), (**d**) low (i.e., $Q < 1 \times 10^7$ ton on 12–15 August), and (**e**) no freshwater discharge (20–22 August). Two profiles on the right side of (**a**,**b**) are the residual over the entire period. The negative values on (**a**,**b**) indicate west- and south-directed residual velocities, respectively.

3.2. Bottom Boundary Layer Flow and Reynolds Stress

In the previous section, we examined the impact of freshwater discharge on the flow of the water column above the bottom boundary layer (BBL). In this section, the response of the bottom boundary layer flow to the rapid seaward flow in the upper layer is examined. Data from the AquaDoppHR, moored in a downward-looking orientation on a tripod during 24–28 August 2011, were used to show the detailed structure of the horizontal and vertical components of the BBL flow, the Reynolds stress components, and the turbulent kinetic energy.

Figure 4 shows the time variation of sea level, the amount of freshwater discharge, and the horizontal and vertical flows at the bottom. The magnitude of flow was about 0.1 m/s in the horizontal component and about 0.01 m/s in the vertical component, and the ebb and flood patterns were not clear since there were several irregular changes in the flow direction even during the flood or ebb tides. The vertical structure of flow indicated that the current magnitude of each profile was similar at 0.15–0.35 mab (Figure 4). However, it began to decrease from 0.15 mab and became almost zero near the bed due to the bottom friction. This phenomenon became clearer as the speed increased. The bottom seaward flow during the low tide was evidence of the fact that the freshwater release can directly affect the bottom flow strength and reverse the flow direction. When freshwater was discharged in the surface layer, the bottom seaward flow was intensified for a short time and then a landward flow appeared (Figure 4b). During 24–28 August 2011, the dam gates opened four times during the low tide: two openings on 24–25 August discharged about 6×10^6 tons of water and the other two openings released over 12×10^6 tons of water (Figure 4a). As the amount of freshwater discharge increased (i.e., on 26 August and 27 August), the change in the flow direction and strength was sharper than the other two cases. One possible reason that the reversed (landward) flow occurred in such a short time could be a result of water mass conservation. When freshwater moved rapidly seaward due to the dam gate opening, it could have produced a return flow in the bottom layer to make the water mass balance inside the estuarine area. The response of the north–south and up–down flow showed a spike-like change just after the freshwater discharge (Figure 4c,d). It is noted, on the other hand, that the flow speed was very weak or almost zero during the no freshwater discharge periods on 25 August and 27 August. It is also interesting to mention the fluctuation of the vertical flow since it displayed directional change (positive to negative) as the freshwater rapidly moved seaward in the upper layer even though the speed was one order of magnitude less than that of the horizontal flow (Figure 4d). The speed of the vertical flow was about 1 cm/s. This fluctuation of the vertical flow occurred during the low tide and was intensified when the freshwater was discharged. Therefore, the freshwater release by opening of the dam gate caused a rapid and strong seaward surface flow, and its impact could reach the bottom and influence the bottom flow structure such as with sudden direction changes in both the horizontal and vertical flow.

At the bottom boundary layer, the increase of near-bed flow can affect the velocity shear and thus intensify the turbulent flow activity. Figure 5 shows the variation of the Reynolds stresses and turbulent kinetic energy in the bottom boundary layer. The vertical structure is similar to the shape of an upside-down bell. The stress terms, $-<u'w'>$, showed a symmetric structure except for a few cases where the values became strongly negative (Figure 5a), and $-<v'w'>$ displayed mostly positive values (Figure 5b). This was related to the fluctuation pattern of the flow. During the freshwater discharge, as noted above, the seaward flow was intensified for a short period of time and then the flow direction was reversed, making $-<v'w'>$ symmetric. However, a mostly southward v-component flow was responsible for the positive values of $-<v'w'>$. Thus, the flow change due to the surface freshwater discharge was linked to the bottom stress intensification and turbulent flow change.

Figure 5c displays the time variation of turbulent kinetic energy. The vertical distribution of TKE is similar to the upside-down half-bell shape and the magnitude rapidly increased from the bed to 0.1–0.2 mab and then remained relatively constant above 0.2 mab. With respect to time, the TKE also increased during the freshwater discharge during low tide, and was proportional to the amount of freshwater discharge. The maximum value during this study period appeared during the late low tide on 27 August.

3.3. Suspended Sediment Concentration and TKE

The major sources of suspended sediment in estuarine environments in general are from the upstream and the sea, as well as bed erosion by flow–bed interaction. In this section, the temporal change in SSC profiles after the freshwater discharges is investigated and the relationship between the turbulent flow activity and SSC in the bottom boundary layer is examined.

Figure 6 illustrates the time variation of SSC profiles during 24–28 August 2011. There were clearly high SSCs found in the surface layer. During the second low tide of each day that freshwater discharge occurred, higher SSCs were observed. It appeared to be due to direct input of suspended sediment discharged from the upstream. As no peaks of SSC were observed when there was no freshwater discharge, for example, during the first low tide of each day, it is evident that the freshwater discharge due to the opening of the dam gate was the main source of the suspended sediment in the surface layer. Another interesting aspect is that high SSC occurred after the freshwater discharge during the flood tide. It is likely that offshore-advected high SSCs during the low tide returned back into the estuary during the flood tide, resulting in high SSCs.

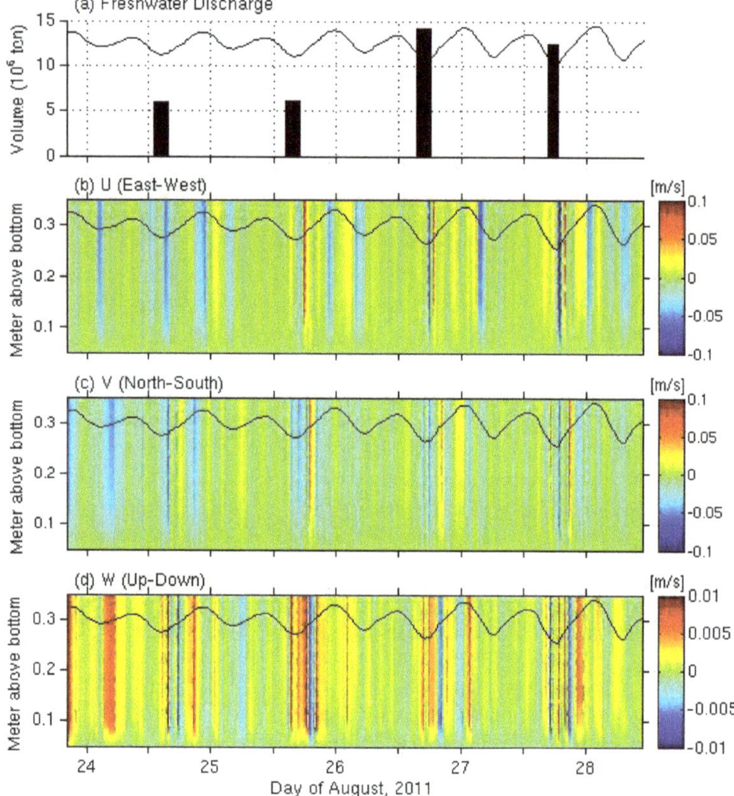

Figure 4. Time variation of the flow structure of the bottom boundary layer with freshwater discharge: (**a**) sea level and freshwater discharge, (**b**) east–west component, (**c**) north–south component, and (**d**) vertical component. The solid line in each panel indicates the sea surface level variation pattern and the scale was adjusted.

Below the surface layer, for example, around 10–12 mab, the SSC was mostly less than 0.01 g/L except during the second flood tide of each day. However, the SSC in the bottom layer suddenly increased and appeared to propagate into the higher layers whenever the surface freshwater was discharged. For example, SSC greater than 0.02 g/L (light green in Figure 6) reached above 6 mab during 25–27 August. In the previous section, we saw the rapid increase of bottom flow speed and stress-related terms (see Figures 4 and 5), and this kind of change in the physical factors could induce erosion of bottom sediments if the bed of this area consisted of very fine sediment such as silt, clay, or mud. According to Kim et al. [24] and Williams et al. [13], the sediments of this estuary consist mainly of silt–clay mixtures in which silt is distributed in the shallow areas and clay exists in the relatively deep areas of the central estuary. Certainly, the bed shear stress observed during the freshwater discharge (see Figure 5) can resuspend the fines high into the water column (Figure 6). Furthermore, Bang et al. [25] simulated sediment transport by using a numerical model, and showed that the discharge-induced estuarine circulation could cause overall silt-size sediment deposition and sustained suspension of clay sediment.

The time series of the TKE and SSC in the bottom boundary layer is illustrated in Figure 7. It is clear that the SSC rapidly increased and then gradually decreased until the next event occurred. The TKE exhibited a peak during each freshwater discharge and was proportional to the amount of freshwater discharge. Likewise, the SSC matched well with the amount of freshwater discharge. This kind of synchronization was repeated during every freshwater discharge. Thus, it could be generalized from this relationship that when the turbulent flow was intensified by the strong seaward surface flow, a relatively large amount of sediment could be resuspended from the bed.

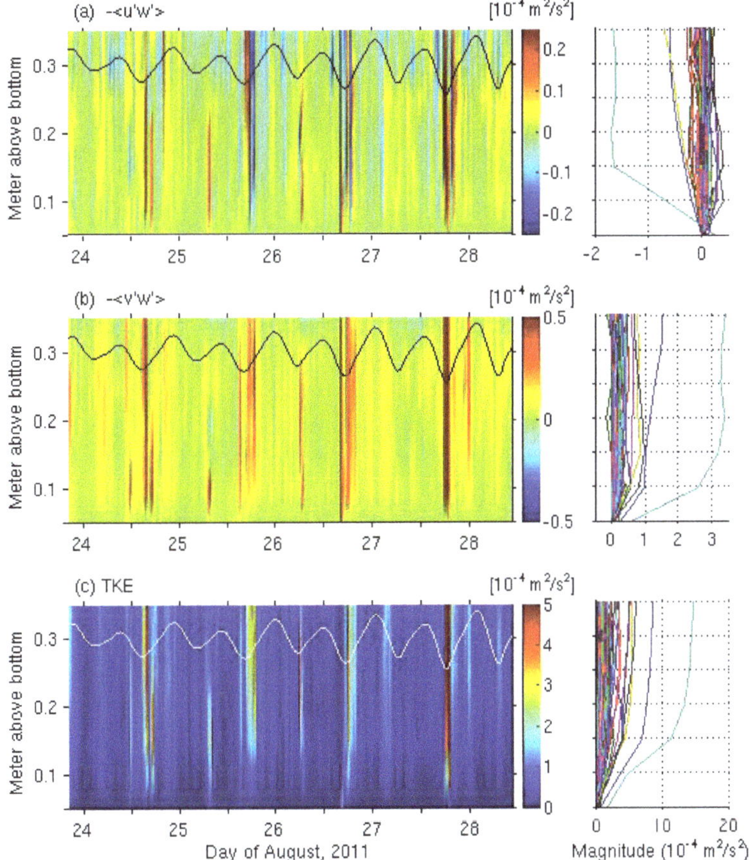

Figure 5. Reynolds stress and turbulent kinetic energy profiles near bed: (**a**) east–west component, (**b**) north–south component, and (**c**) turbulent kinetic energy (TKE). Black and white solid lines in the left panel indicate the sea level variation.

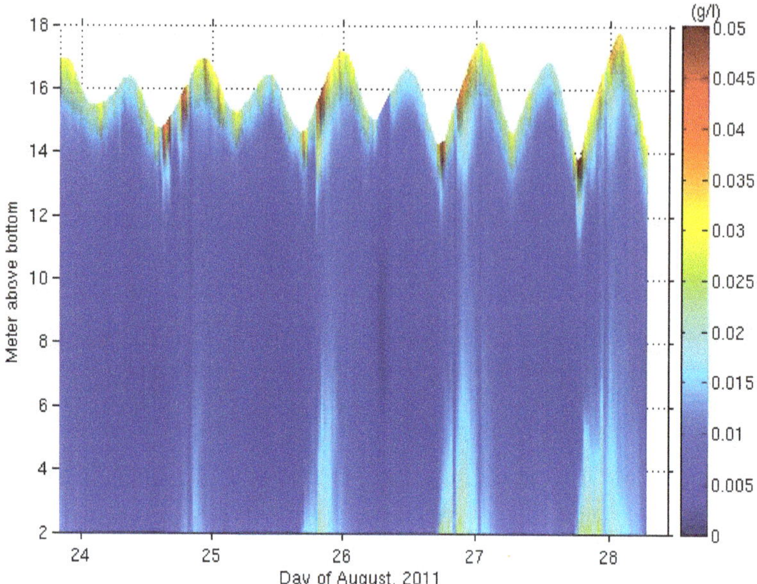

Figure 6. Variation of suspended sediment concentration during 24–28 August 2011.

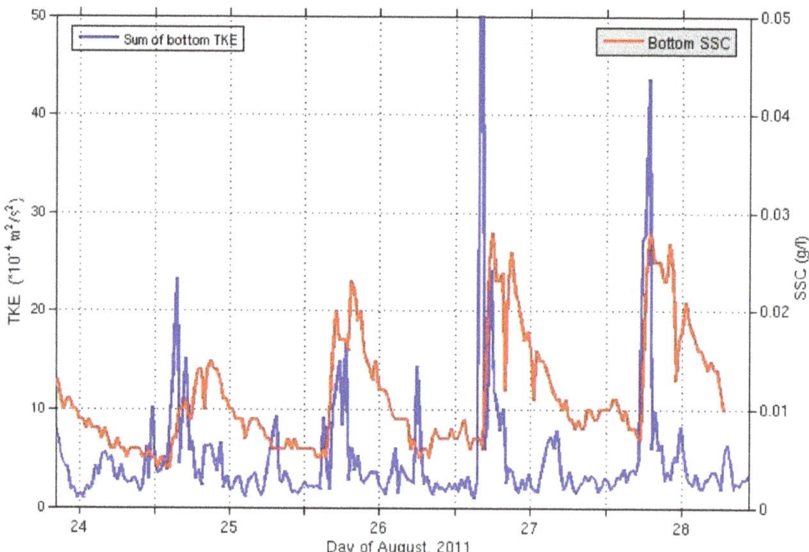

Figure 7. Comparison between the turbulent kinetic energy (TKE) and suspended sediment concentration (SSC) changes in the bottom boundary layer.

4. Discussion

In estuaries altered by an estuarine dam, the opening of the dam gates results in a sudden release of a significant amount of freshwater to the river mouth area, resulting in changes to the physical and environmental conditions such as the formation of a strong surface current and sediment suspension. Measurement of the water currents during August 2011 in the Yeongsan estuary was carried out to investigate the changes in the vertical flow structure, the bottom turbulent flow, and the relationship between the bottom turbulent flow activity and the variation of suspended sediment concentration during the freshwater discharge.

The freshwater discharge due to dam gate opening could significantly affect the vertical structure of flow, inducing velocity shear and turbulent flow activity in the bottom boundary layer. During the period of no or weak discharge, the flow was mostly tidal motion and the vertical structure showed multiple layers. Cho et al. [11] showed that the spreading of warm freshwater discharged over the pre-existing surface water and the intrusion of warm saline water along the bottom from the open sea produced a multilayer structure, and the multilayer structure remained throughout the summer due to strong stratification and a weak tidal current in the Yeongsan estuary. During the freshwater discharge, however, a two-layer system developed with seaward flow in the surface and landward flow in the lower layer [7,26]. The thickness of the seaward flow was affected by the amount of freshwater, and the thickness of landward flow in the subsurface layer increased as the freshwater amount decreased. The magnitude of the residual flow was about 10 cm/s in the surface during the discharge periods, and became stagnant with weak, vertically variable horizontal flows during no discharge periods.

The bottom flow near the bed did not follow the general flood and ebb cycle in tidal motion. The seaward and landward flows were, however, formed after freshwater discharge, for example, on 25–27 August. This appears as strong evidence that the freshwater release due to opening of the dam gate affected the bottom flow strength. One more interesting aspect is that the reversed (landward) flow followed after the initial seaward flow. A two-layer circulation system was suggested wherein, following the water mass conservation principle, a landward flow could happen in order to recover the water overtransported to the sea by the suddenly intensified seaward flow which could cause additional seaward movement. Considering the concomitant vertical flow pattern of up and down, the upward flow above the bed was first shown when freshwater was released and then downward flow happened with a maximum magnitude of 0.01 m/s.

An impact of surface freshwater discharge to the bottom boundary layer was an intensification of the turbulent flow activity. As the bottom flow changed because of the surface freshwater discharge, the fluctuation of seaward and landward flow over a short period of time caused the stress to increase in the bottom boundary layer which was linked to an increase in TKE. The Reynolds stresses showed a symmetric structure in their vertical distribution and their magnitude rapidly increased during the freshwater discharge period with a maximum magnitude of 2×10^{-4} m^2/s^2. The TKE structure had the shape of an upside-down half-bell with an increase from the bed to 0.1–0.2 mab. Like the stresses, the magnitude of TKE was proportional to the amount of freshwater discharge and reached over 1×10^{-3} m^2/s^2 for the discharged period.

The SSC estimated from the acoustic backscatter signal of the four beams of the ADCP displayed high values in the surface and bottom layers. The high SSC in the surface happened after or during the freshwater discharge, indicating that the main source of suspended sediment was the upstream river water because there was not any other source of sediment during low tide and there was no such peak value of SSC when there was no freshwater discharged. The rapid increase of SSC in the bottom also happened after the surface freshwater discharge and it gradually propagated to the upper layer, and since it happened at the same time or after the sudden increase of the bottom TKE, the SSC near the bed was related to the intensified bottom turbulent flow that was the result of surface freshwater discharge due to the opening of the dam gate.

5. Conclusions

Using ADCP measurements in the altered Yeongsan estuary, we examined the flow in the water column and above the bed resulting from opening of the dam gate and the release of water with different properties (salinity, temperature, flow rate). The freshwater discharge was responsible for intensifying the two-layer circulation of offshore surface flow and landward bottom flow. The rapid and strong seaward surface flow affected the bottom flow structure, leading to sudden directional changes in both the horizontal and vertical flow. Responding to the freshwater discharge, the bottom turbulence also intensified rapidly, which in turn resuspended a large amount of sediment from the bed. The results of this study indicate that the surface freshwater discharge due to opening of the estuarine dam gate affects the behavior of water flow, bottom turbulence, and sediment transport in the altered Yeongsan estuary. Finally, it should be noted that since many estuarine dams have been constructed rapidly all over the world, for example, in the estuaries of the Senegal River, the Rhine-Muse rivers, and the Murray-Darling rivers, and that the construction of new estuarine dams is also under consideration, the results of this study could provide valuable insight into morphological change in estuarine environments with man-induced discrete freshwater discharges.

Author Contributions: Conceptualization, methodology, resources, writing—review and editing, G.-h.L.; formal analysis, investigation, validation, writing—original draft preparation, K.K. All authors have read and agreed to the published version of the manuscript.

Funding: This research was funded by Inha University Research Grant, 63142-01" and by the National Institute of Meteorological Sciences of the Korea Meteorological Administration project titled "Development of Marine Meteorology Monitoring and next-generation Ocean Forecasting System (KMA2018-00420).

Acknowledgments: The authors would like to thank anonymous reviewers for their constructive comments to make the high quality research.

Conflicts of Interest: The authors declare no conflict of interest.

References

1. Dyer, R. *Estuaries: A Physical Introduction*, 2nd ed.; John Wiley & Sons Ltd.: Chichester, UK, 1997; pp. 1–32.
2. Lee, G.; Dade, W.B.; Friedrichs, C.T.; Vincent, C.E. Spectral estimates of bed shear stress using suspended-sediment concentrations in a wave-current boundary layer. *J. Geophys. Res.* **2003**, *108*, 1–15. [CrossRef]
3. Holmedal, L.E.; Myrhaug, D. Boundary layer flow and net sediment transport beneath asymmetrical waves. *Cont. Shelf Res.* **2006**, *26*, 252–268. [CrossRef]
4. Wren, D.G.; Kuhnle, R.A.; Wilson, C.G. Measurements of the relationship between turbulence and sediment in suspension over mobile sand dunes in a laboratory flume. *J. Geophys. Res. Earth Surf.* **2007**, *112*, 1–14. [CrossRef]
5. Byun, D.S.; Wang, X.H.; Holloway, P.E. Tidal characteristic adjustment due to dyke and seawall construction in the Mokpo Coastal Zone, Korea. *Estuar. Coast. Shelf Sci.* **2004**, *59*, 185–196. [CrossRef]
6. Kang, J.W. Changes in tidal characteristics as a result of the construction of sea-dike/sea-walls in the Mokpo coastal zone in Korea. *Estuar. Coast. Shelf Sci.* **1999**, *48*, 429–438. [CrossRef]
7. Shin, H.J.; Lee, G.; Kang, K.; Park, K. Shift of estuarine type in altered estuaries. *Anthr. Coasts* **2019**, *170*, 145–170. [CrossRef]
8. Jung, T.S.; Kim, T.S. Prediction System of Hydrodynamic Circulation and Freshwater Dispersion in Mokpo Coastal Zone. *J. Korean Soc. Mar. Environ. Energy* **2008**, *11*, 13–23.
9. Kang, J.W.; Kim, Y.S.; Park, S.J.; So, J.K. 3-D Applicability of the ESCORT Model—Simulation of Freshwater Discharge. *J. Korean Soc. Coast. Ocean Eng.* **2009**, *21*, 230–240.
10. Park, L.H.; Cho, Y.K.; Cho, C.; Sun, Y.J.; Park, K.Y. Hydrography and circulation in the Youngsan river estuary in summer, 2000. *J. Korean Soc. Oceanogr.* **2001**, *6*, 218–224.
11. Cho, Y.; Park, L.; Cho, C.; Tae, I.; Park, K. Multi-layer structure in the Youngsan Estuary, Korea. *Estuar. Coast. Shelf Sci.* **2004**, *61*, 325–329. [CrossRef]
12. Park, H.B.; Kang, K.; Lee, G.; Shin, H.J. Distribution of Salinity and Temperature due to the Freshwater Discharge in the Yeongsan Estuary in the Summer of 2010. *J. Korean Soc. Oceanogr.* **2012**, *17*, 139–148.
13. Williams, J.; Dellapenna, T.; Lee, G.; Louchouarn, P. Sedimentary impacts of anthropogenic alterations on the Yeongsan Estuary, South Korea. *Mar. Geol.* **2014**, *357*, 256–271. [CrossRef]
14. Dahm, W.J.A.; Dimotakis, P.E. Measurements of entrainment and mixing in turbulent jets. *AIAA J.* **1987**, *25*, 1216–1223. [CrossRef]
15. Van Rhee, C. Sediment Entrainment at High Flow Velocity. *J. Hydraul. Eng.* **2010**, *136*, 572–583. [CrossRef]
16. Hoffmans, G.J.C.M. Jet scour in equilibrium phase. *J. Hydraul. Eng.* **1998**, *124*, 430–437. [CrossRef]

17. Althaus, J.M.I.J.; De Cesare, G.; Schleiss, A.J. Fine sediment release from a reservoir by controlled hydrodynamic mixing Measurement devices. In Proceedings of the 34th World Congress of the International Association for Hydro-Environment Research and Engineering, Brisbane, Australia, 26 June–1 July 2011; pp. 1763–1770.

18. Althaus, J.M.I.J.; Cesare, G.D.; Schleiss, A.J. Sediment Evacuation from Reservoirs through Intakes by Jet-Induced Flow. *J. Hydraul. Eng.* **2015**, *141*, 04014078. [CrossRef]

19. Wohl, E.; Cenderelli, A. Sediment deposition and transport patterns following a reservoir sediment release. *Water Resour. Res.* **2000**, *36*, 319–333. [CrossRef]

20. Foreman, M. *Manual for Tidal Currents Analysis and Prediction*; Institute of Ocean Sciences, Patricia Bay: Sidney, BC, Canada, 2004; pp. 2–30.

21. Pawlowicz, R.; Beardsley, B.; Lentz, S. Classical tidal harmonic analysis including error estimates in MATLAB using T TIDE. *Comput. Geosci.* **2002**, *28*, 929–937. [CrossRef]

22. Park, H.B.; Lee, G. Evaluation of ADCP backscatter inversion to suspended sediment concentration in estuarine environments. *Ocean Sci. J.* **2016**, *51*, 109–125. [CrossRef]

23. Hansen, D.V.; Rattray, M. Gravitational Circulation in Straits and Estuaries. *J. Mar. Res.* **1966**, *23*, 104–122.

24. Kim, Y.; Chang, J.H. Long-term Changes of Bathymetry and Surface Sediments in the dammed Yeongsan River Estuary, Korea, and Their Depositional Implication. *Sea* **2017**, *22*, 88–102.

25. Bang, K.; Kim, T.I.; Song, Y.S.; Lee, J.H.; Kim, S.W.; Cho, J.; Kim, J.W.; Woo, S.B.; Oh, J.K. Numerical Modeling of Sediment Transport during the 2011 Summer Flood in the Youngsan River Estuary, Korea. *J. Korean Soc. Coast. Ocean Eng.* **2013**, *25*, 76–93. [CrossRef]

26. Kim, Y.H.; Hong, S.; Song, Y.S.; Lee, H.; Kim, H.; Ryu, J.; Park, J.; Kwon, B.; Lee, C.; Khim, J.S. Seasonal variability of estuarine dynamics due to freshwater discharge and its influence on biological productivity in Yeongsan River Estuary, Korea. *Chemosphere* **2017**, *181*, 390–399. [CrossRef] [PubMed]

Article

Turbulence Characteristics before and after Scour Upstream of a Scaled-Down Bridge Pier Model

Seung Oh Lee [1] and Seung Ho Hong [2],*

[1]   Department of Civil Engineering, Hongik University, 94 Wausan-ro, Mapo-gu, Seoul 04066, Korea
[2]   Department of Civil and Environmental Engineering, West Virginia University, 1306 Evansdale Drive, Morgantown, WV 26506, USA
*   Correspondence: sehong@mail.wvu.edu

Received: 19 August 2019; Accepted: 9 September 2019; Published: 12 September 2019

Abstract: Bridge pier scour is one of the main causes of bridge failure and a major factor that contributes to the total construction and maintenance costs of bridge. Recently, because of unexpected high water during extreme hydrologic events, the resilience and security of hydraulic infrastructure with respect to the scour protection measure along a river reach has become a more immediate topic for river engineering society. Although numerous studies have been conducted to suggest pier scour estimation formulas, understanding of turbulence characteristics which is dominant driver of sediment transport around a pier foundation is still questionable. Thus, to understand near bed turbulence characteristics and resulting sediment transport around a pier, hydraulic laboratory experiments were conducted in a prismatic rectangular flume using scale-down bridge pier models. Three-dimensional velocities and turbulent intensities before and after scour were measured with Acoustic Doppler Velocimeter (ADV), and the results were compared/analyzed using the best available tools and current knowledge gained from recent studies. The results show that the mean flow variable is not enough to explain complex turbulent flow field around the pier leading to the maximum scour because of unsteady flows. Furthermore, results of quadrant analysis of velocity measurements just upstream of the pier in the horseshoe vortex region show significant differences before and after scour.

Keywords: bridge pier; horseshoe vortex; Physical hydraulic modeling; quadrant analysis; Scour and Velocity field

---

1. Introduction

Bridge pier scour is one type of local scour caused by sediment transport that is driven by local flow structure; therefore, it is necessary to be acquainted with the flow structure and the related scour mechanisms around the bridge pier. In general, the local flow structure around a bridge pier is composed of downflow at the upstream face of the pier in the vertical plane; the horseshoe vortex system that wraps around the base of the pier, which is the primary contributor to local scour upstream of the pier; the bow wave near the free surface on the upstream face of the pier; and the wake vortex system at the rear of the bridge pier that extends over the flow depth [1–3]. These features greatly complicate the understanding of the local flow structures [4,5], and the comprehensive effect of those complex flow structures is to increase the local sediment transport leading to additional local scour around the bridge pier [6,7].

When flow approaches a bridge pier, the velocity becomes zero on the upstream face of the bridge pier. Due to the strong adverse pressure gradient imposed by the bridge pier in the streamwise approach flow direction, the boundary layer separates upstream of the bridge pier. In the separated region, several vortices are consecutively developed, and subsequently stretched around the base of the bridge pier, giving rise to what is called a horseshoe vortex system. The primary horseshoe vortices rotate in the same sense as the approach boundary layer vorticity, but secondary vortices have the opposite rotation to preserve streamline topology. During the time that the horseshoe vortex nearest the bridge pier is decreasing in size due to stretching, a newer and younger secondary separation vortex is induced upstream of the primary vortex. The size and strength of the secondary horseshoe vortex increases with time while the size of the primary vortex continues to be reduced by stretching. At some time, the secondary and smaller vortices merge with the primary horseshoe vortex or leapfrog it to strengthen the primary horseshoe vortex, which is finally stretched completely around the bridge pier temporarily stabilizing the flow. Subsequently, instability occurs and the primary vortex forms again. The process is then repeated irregularly [8,9].

In addition to the theoretical descriptions of local vortex structures, to find more general relationships between the vortex system around a bridge pier and the resulting sediment transport, many researchers have adopted analytical, experimental, as well as computational approaches. Melville [10] has observed that the size and circulation of the horseshoe vortex increases rapidly, and the velocity near the bottom of the hole decreases as the scour hole is enlarged. According to Melville, magnitude of the downflow seems to be directly associated with the rate of scour. However, Baker [11] argued that because the size of the scour hole is much larger than the size of the vortex core on a flat-bed before scouring, such a calculation would predict wrongly that the circulation increases as scour proceeds. In his study, Baker [11] used the horseshoe vortex core circulation to derive an equation to predict the scour depth. Baker [11] assumed that the horseshoe vortex strength in the scour hole can be equal to that on a flat-bed as scour depth develops and suggested the equation with respect to pier width and free stream velocity. Later, Nakagawa and Suzuki [12] assumed that the scale and strength of the primary horseshoe vortex are constant during evolution of the local scour hole and suggested a scour prediction equation with using the stochastic nature of particle movement, originally developed by Einstein [13]. At similar time, Qadar [14] experimentally hypothesized the maximum scour depth to be a function of the initial vortex strength which is composed of vortex size and stream velocity. Based on the experimental results, an envelope curve was proposed to show the relationship between scour depth and initial vortex strength. Qadar [14] quantitatively evaluated his results by comparing with laboratory and field data.

After Qadar [14], Kothyari et al. [15] explored the diameter of primary vortex using a regression analysis of experimental data on flow around a cylindrical pier. They mentioned that the diameter of the primary vortex is dependent on the bridge opening width in comparison to the size of pier diameter. Also, Ram [16] expressed an equation for initial diameter of the horseshoe vortex and, in his findings, the initial diameter of the vortex decreases with increasing the pier Reynolds number (Re_b) represented by a pier width as a characteristics length. However, Muzzammil and Gangadhariah [17] found that the relative vortex size, which is the ratio of vortex diameter and pier width, is weakly influenced by the pier Reynolds number for higher values on a rigid flat bed. This means that vortex size is only dependent on the pier width for higher values of the pier Reynolds number ($Re_b > 10^4$). Based on analytical models relating scour depth to horseshoe vortex properties, Muzzammil and Gangadhariah [17] proposed that equilibrium scour depth is a function of the horseshoe vortex size, tangential velocity, and vortex strength in the scour hole. They found that the mean size of the horseshoe vortex is ~20% of the cylindrical pier diameter, and the vortex tangential velocity is ~50% of the mean velocity of approach flow for $10^4 \leq Re_b \leq 1.4 \times 10^5$ at fixed flat-bed conditions. The size of the vortex is assumed to be independent of the sediment mobility.

More recently, Dey and Raikar [7] found that the flow and turbulence intensities in the horseshoe vortex in a developing scour hole are reasonably comparable with those in before scour. Similar results can be found in other studies conducted with large-eddy simulations (LES) and detached-eddy simulation (DES) [18–20]. However, based on the experimental studies using particle image velocimetry (PIV), Guan et al. [21] argued that the size of the main vortex responsible for the maximum scour depth upstream of the pier increases with increasing scour depth.

In clear-water scour regime, as the scour depth increases, the shear stress beneath the horseshoe vortex is reduced until the shear stress becomes less than the critical shear stress and sediment movement ceases in the scour hole. As explained in the previous paragraphs, even if the role, size, and strength of turbulence leading to pier scour have been investigated qualitatively, the coherent turbulent characteristics and associated bursting events for sediment transport is still an active area of interest. So far, several of methods presented in literatures for describing the horseshoe vortex properties have not considered turbulent characteristics or unsteadiness of the horseshoe vortex; but, in fact, the bed materials around a bridge pier move irregularly in time even if the approach flow is steady. It is found in the literature that the strength and size of the horseshoe vortex are closely related to the pier geometry and the approach flow velocity upstream of the bridge while the scour depth cannot be accurately predicted unless a clear relationship between the large-scale unsteadiness of the horseshoe vortex and sediment size is presented quantitatively. One of the important objectives of this study is investigating turbulence characteristics and sequential occurrence of bursting turbulence events before and after scour in the process of particle entrainment in front of a pier. Thus, to comprehensively attack the objectives, hydraulic laboratory experiments were conducted in a flume using two different scaled-down bridge pier models. Visual observations by high speed camera as well as three-dimensional velocities and turbulent intensities before and after scour measured with ADV were analyzed using the best available tools and current knowledge.

2. Methodology

2.1. Experimental Setup

As shown in Figure 1a,b, in previous studies [22–24], laboratory experiments were conducted using various scaled hydraulic model of the Chattahoochee River bridge at Cornelia, Georgia, and Flint River bridge at Bainbridge, Georgia, USA, respectively, including the full river bathymetry. The previous experimental studies successfully explored the effect of sediment size on pier scour depth at different geometric model scales and, also based on the large number of experimental and field investigations/comparisons, improved local scour formulas were suggested. Furthermore, strategies of deciding sediment size for scaled-down hydraulic modeling was also proposed. After those model studies, each pier bent were carefully removed from the laboratory flume. Figure 1c,d show example of bridge pier bent model of Chattahoochee River bridge and Flint River bridge, respectively, with individual scales for this study. Pier bent consists of four rectangular concrete columns and rectangular concrete footings for Chattahoochee River bridge, as shown in Figure 1c. However, for Flint River bridge, two square concrete pier columns are placed on large stepped square concrete footings.

The removed pier bent model shown in Figure 1c,d was re-built inside a 1.1 m wide by 24.4 m long glass-sided tilting flume to investigate detailed turbulence and flow characteristics in front of the pier before and after scour. The approach section of the pier model was 15.0 m long followed by a working mobile bed section with a length of 3.0 m in which the pier model was placed. The length of the approach section was decided based on the findings from other researches [25–29] to ensure a fully developed approach turbulent flow and turbulent boundary layer at the bridge pier. After the working mobile bed section, additional 3 m long sediment trap section downstream of the pier model was placed. For the current set of experiments, instead of using actual cross section shape and river geometry used in the previous experiments, the channel was constructed to have a straight alignment rather than meandering, and rectangular shape of cross section was maintained through the entire flume to find more general features of flow and turbulence fields. The flume was filled with 0.53 mm bed sediment for each experiments and carefully leveled to the elevations established by the field measurements in each site. Figure 2a, b show experimental setup before running scour experiment for Chattahoochee River bridge model and Flint River bridge model, respectively.

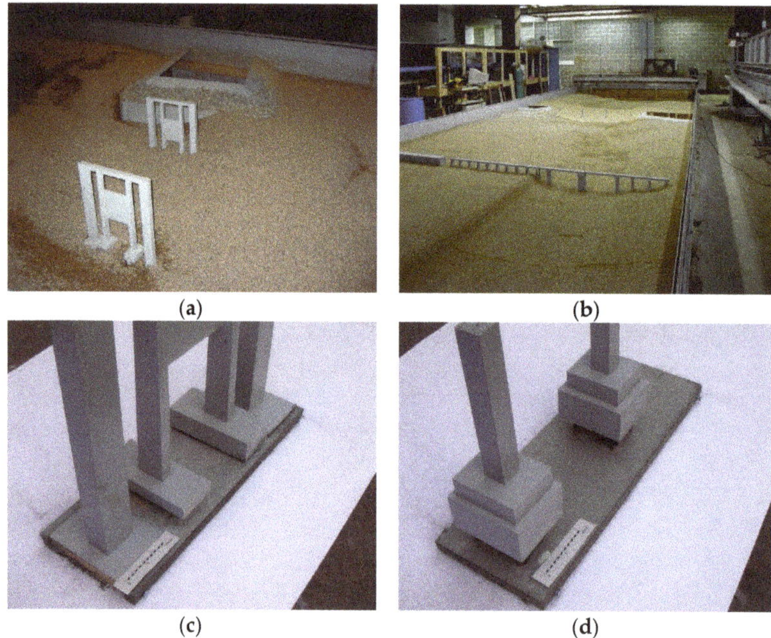

Figure 1. Hydraulic laboratory model in previous laboratory studies and bridge pier model for this study: (**a,c**) 1:23 scaled model for Chattahoochee River bridge and (**b,d**) 1:33 scaled model for Flint River bridge.

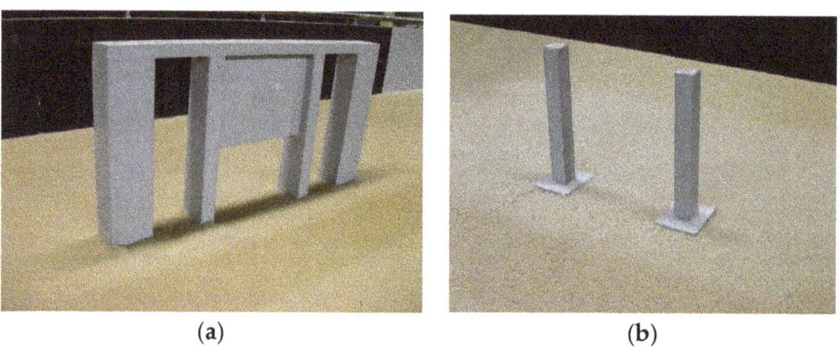

Figure 2. Pier model in the flume: (**a**) Chattahoochee River bridge model and (**b**) Flint River bridge model.

2.2. Experimental Procedure

The experimental campaign consists of two scenarios: moveable bed experiment and fixed-bed experiment. First, moveable bed experiments were conducted to investigate variance of hydraulic parameters affecting scour depth over time. The flume was slowly filled with water to saturate the sand. After complete saturation, the required discharge (uncertainty of $\pm 2.8 \times 10^{-4}$ m^3/s) was established with the flow depth adjusted well above the desired value. Then, the flow depth was gradually decreased by changing the height of the tailgate until the target approach flow depth was obtained. During this time, the point gage (uncertainty of ± 1 *mm*) was used to monitor the flow depth. Once the target flowrate and the flow depth had been reached, scour continued for 2 to 3 days until equilibrium was achieved. The equilibrium was defined when the increment of scour depth is less than 5% of the bride pier diameter during 24 h. During the scouring process, instantaneous point velocities and turbulence quantities were measured by ADV in front of the pier. Furthermore, temporal change of bed elevations were measured periodically using ADV temporarily positioned for a moment above the point of scouring. At the end of scouring (equilibrium state), the velocity flow field was measured throughout the test section both in the near field next to the pier and in the far field at relative elevations for the comparison of turbulence and flow characteristics between after scour and before scour. During the velocity measurements, ADV sampling frequency was chosen to be 25 Hz with a duration of at least 2 min and perhaps as much as five minutes depending on the magnitude of turbulence at each measuring location. The correlation values in these measurements were greater than 80% and the Signal Noise Ratio (SNR) was greater than 15. The phase-space despiking algorithm was also employed to remove any spikes in the time record caused by aliasing of the Doppler signal which sometimes occurs near a boundary. More detailed filtering protocol can be found in Lee and Sturm [22] and Hong et al. [26]. After the completion of each experiment, the final bed elevations were measured using the ADV and the point gage.

When the moveable bed experiments were completed, the entire moveable bed was re-leveled and fixed by spraying polyurethane. During fixed-bed experiments, the same flow conditions as those in moveable bed experiments were reproduced, and velocities and turbulence quantities were measured in the same way as in the moveable bed experiments to investigate the effect of initial flow parameters responsible for the scour. In addition to the measurements of initial flow parameters, flow visualization experiments were also conducted. A kaolinite suspension was used as a tracer upstream of the bridge pier to show the flow structure around the bridge pier and also to detect the frequency of the horseshoe vortex system immediately upstream of the pier. The tracer was transferred from a conical tank by an electric pump operating at a maximum flowrate of 0.003 m^3/s. The kaolinite suspension was mixed to achieve a concentration of 1.0 mg/cm^3 in the tank. The flow rate of tracer was adjusted with the aid of rotameter to produce a released velocity that was the same as the open channel mean velocity. As the tracer was released at a constant rate, a high speed video camera (30 FPS) was used to capture the unsteady dynamics of the swirl of the horseshoe vortex as it amplified and partially collapsed in size.

3. Results and Discussion

The experimental conditions have been summarized in Table 1, where Q is the total discharge, b is the width of bridge pier, y_1 is approach section water depth, V_1 is approach section velocity, T_{eq} is time to the equilibrium scour, and d_s is the equilibrium scour depth in front of the bridge pier.

Table 1. Summary of measured experimental conditions.

Run	Model	Scale	Q (m^3/s)	b (m)	y_1 (m)	V_1 (m/s)	T_{eq} (h)	d_s (m)	Conditions
1	CR	1:23	0.051	0.046	0.191	0.257	30	0.093	Fixed &
2			0.044	0.046	0.142	0.304	12	0.090	Moveable-bed
3	FR	1:33	0.054	0.055	0.241	0.215	48	0.046	Fixed &
4			0.052	0.055	0.170	0.281	24	0.085	Moveable-bed

CR: Chattahoochee River bridge model and FR: Flint River bridge model.

3.1. Velocity Field

The velocity fields around the pier bent were measured for the Chattahoochee River model and Flint River model in both the fixed-bed (before scouring) and moveable-bed (after scouring in equilibrium). Figure 3a,c shows the velocity fields of "before scour" for Run 1 and Run 3, respectively. The longitudinal distance (x) and lateral distance (y) were normalized with width of corresponding bridge pier model and the near field vectors measured at 40 percent of the approach flow depth were normalized by measured approach velocity. Higher velocities were shown on both sides of the first pier, where deeper scour occurred as the flow curved around the pier bent. In the wake zone, the mean velocities became smaller than those in the outer region. The velocity defected in this region gradually recovered in the downstream direction. The magnitude of mean velocities upstream of the first pier along the centerline became smaller approaching the pier stagnation line due to the existence of the pier. Figure 3b,d shows the combination of scour depth contours and the near-field velocities measured at 40 percent of the approach flow depth under the same flow condition as in "before scouring" but at the completion of scour for Run 1 and Run 3, respectively. The near-field velocity distributions were very close to being symmetric with respect to the centerline of the bridge pier bent as shown in Figure 3 for both runs. The characteristic decrease in magnitude was observed in near field velocity around the pier bents when comparing results before and after scour. The maximum relative difference in magnitude was approximately 30–40 percent for both cases in the vicinity of the first pier on the right-hand side. Interestingly, the maximum scour depth occurred at the nose of front pier for Run 1, as expected, however for the case of Run 3, the maximum scour occurred between the two piers with a high degree of symmetry on the left and right sides as shown in Figure 3d. For the Flint River bridge pier bent, the pier columns are placed on large stepped footings and the footings are already exposed at the beginning of scour as shown in Figure 2b. Because the footing intercepts the downflow along the nose of pier bent which feeds the size and strength of horseshoe vortex, the amount of local scour depth in front of the first pier was reduced [30–32].

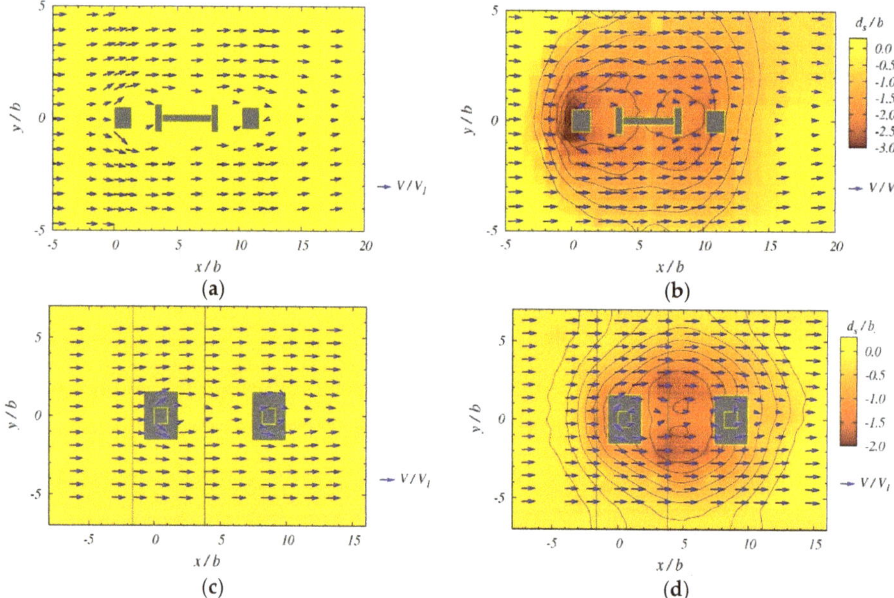

Figure 3. Normalized mean velocity vectors: Measured at 40 percent of the approach depth for experimental Run 1 before and after scour in (**a**,**b**), respectively, and Run 3 before and after scour in (**c**,**d**), respectively.

3.2. Temporal Variation of Flow and Turbulence Characteristics Upstream of the Bridge Pier

The measured mean velocity and turbulence kinetic energy (TKE) fields have been usually used to validate three-dimensional numerical models. Even though the simulated velocity profiles at several locations were in good agreement compared with the laboratory experiments, quantitative connections with the scour depth are difficult to make. In addition, results from a three-dimensional numerical model show that the maximum mean shear stress on a fixed bed does not correspond with the maximum depth of scour hole in front of the piers [33]. Furthermore, as shown in Figure 3, the horizontal mean (time-averaged) point velocity vector plots did not indicate large changes in the velocity field with scour development when comparing the before- and after-scour conditions. It is concluded that the details of the horseshoe vortex itself must be investigated further to understand the development of the scour hole in front of the pier rather than the general-mean turbulence and flow characteristics of the near field.

Thus, in an effort to better understand the relationship between the flow field and the resulting pier scour over the development of scour hole, bed elevations, and three-dimensional velocity components as well as turbulence intensities upstream of the bridge pier were measured intermittently at two points during the scour process to capture the temporal variation of flow and turbulence characteristics as the scour hole developed. The flow was continued for the measurements so that the measurements could be completed in a time duration that was short (approximately 2–3 min) in comparison to the rate of scour hole development. The two measured points were located horizontally at a distance of one pier width upstream of the bridge pier in the streamwise direction because the size of horseshoe vortex is comparable with the size of pier width. As shown in Figure 4, the vertical location of one point (*LOC1*) was in the scour hole itself where it was varied to maintain a constant vertical displacement above the bed as the scour hole deepened with time. The other point (*LOC2*) was fixed above *LOC1* but at a constant elevation that was close to the initial bed elevation before scour began as also shown in Figure 4. Sufficient clearance between the bridge pier and the ADV probe was required in order to place the ADV probe without bumping the pier or disturbing the horseshoe vortex system upstream of the bridge pier. Thus, the Chattahoochee River bridge model was chosen for this experiment based on the physical size and simple footing shape compared to the Flint River model.

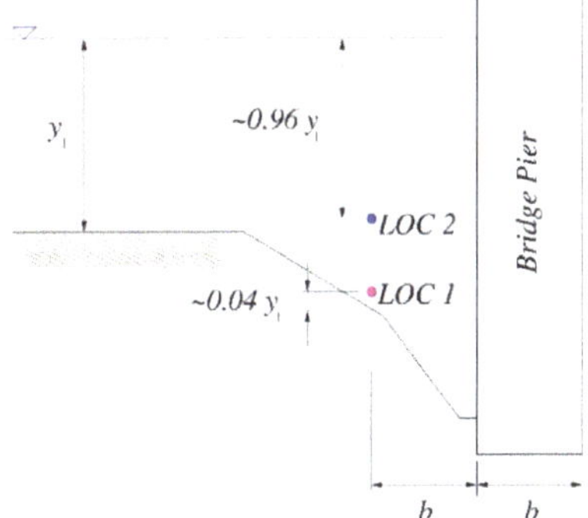

Figure 4. Schematic diagram of the locations for measuring the temporal variation of flow characteristics.

The streamwise (*U*) and vertical (*W*) time-averaged velocity profiles normalized by approach mean velocity at *LOC1* and *LOC2* are shown in Figure 5. Both velocity profiles at *LOC1* fluctuated slightly with time, but the values remained close to zero as the scour depth increased over time which is shown on the secondary vertical axis in Figure 5a. The averaged values became close to zero at *LOC1* since it is located near the bed in the separation zone. However, the velocity profiles at *LOC2* fluctuated significantly with time as shown in Figure 5b because the vertical position of *LOC2* moved into the highly turbulent region associated with the horseshoe vortex during development of the scour hole. The fluctuations in the streamwise velocity seem to be associated with the intermittent fluctuations in the scour hole depth that result from the collapse of the sides of the hole, followed by further scouring as the hole enlarges.

The turbulence intensity in both the streamwise (*u′*) and vertical (*w′*) directions at *LOC1* increased during the initial stage of scour development and then became smaller with time as shown in Figure 6a. The large fluctuations of streamwise turbulence intensity during the initial stages of scour development is due to the unsteadiness of the location of the separation point upstream of the pier. As the scour depth changed rapidly during the initial stage, the vertical turbulence intensity became approximately four times larger than at later stages of scour hole development, indicating that the contribution of vertical turbulence intensity to the rate of scour development was very significant during the initial stage. Figure 6b shows that the turbulence intensity in both vertical and streamwise directions at *LOC2* approached the same constant value as the scour hole developed. The vertical turbulence intensity at *LOC2* was relatively small at the beginning of the scour process, but then increased significantly to a value greater than that at *LOC1*. This is due to the movement of the vertical location of *LOC2* into the highly turbulent region associated with the horseshoe vortex.

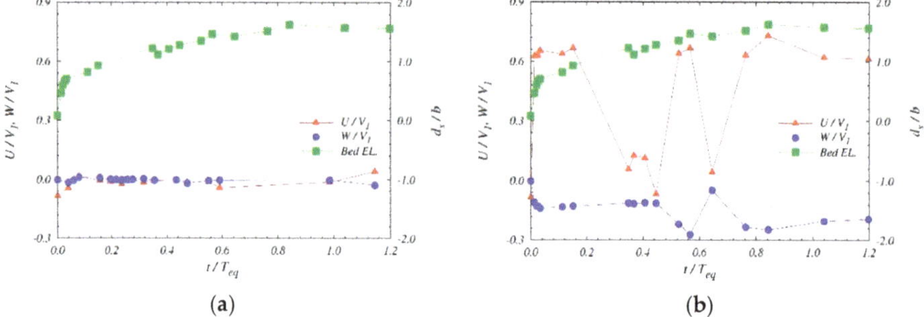

Figure 5. Velocity profile over time at (**a**) *LOC* 1 and (**b**) *LOC* 2 for Run 1.

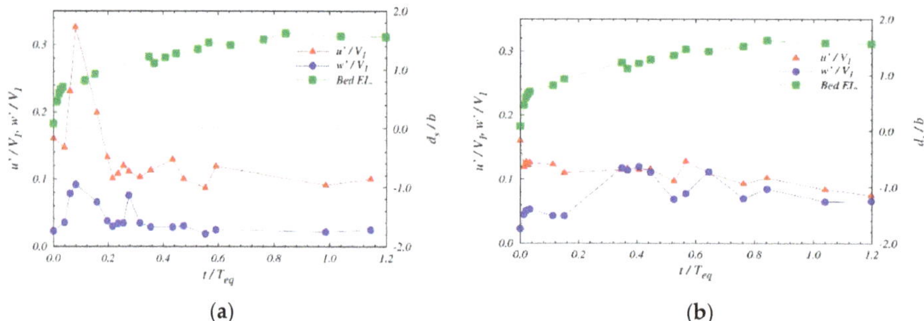

Figure 6. Temporal variations of turbulence intensity profile at (**a**) *LOC* 1 and (**b**) *LOC* 2 for Run 1.

3.3. Flow Characteristics Upstream of the Bridge Pier

While in the previous section, the long-term temporal development was considered in relationship to the variance of velocity and turbulence characteristics in the scour hole, this section investigates the short-term transient behavior of the flow immediately upstream of the bridge pier on a fixed flat bed in the region of flow separation and the horseshoe vortex. The primary horseshoe vortex is unsteady with the formation of a system of secondary vortices, and the secondary vortices are quasi-periodically combined with the primary vortex increasing its size and strength depending on the degree of stretching around the pier [8,9]. Thus, the primary horseshoe vortex oscillates in position and size in an irregular shift between two modes of behavior which are expanding and contracting over time. Those oscillation of horseshoe vortex in its size and strength in front of the pier was captured by high-speed camera in Figure 7 in this experiment, and the time difference (Δ*t*) between Figure 7a,d was related to the frequency of horseshoe vortex [22,34]. As shown in Figure 7, because of those two modes, the instantaneous velocity time series when measured near the bed close to the pier alternately exhibits periods of positive streamwise velocity towards the pier followed by negative streamwise velocity away from the pier. The result is a bimodal velocity distribution, first described by Devenport and Simpson [34], and further studied experimentally and numerically [18,22,35]. Those two alternate states of the horseshoe vortex are bistable with the contracted mode occurring approximately 20–30% of the time. Furthermore, they found that the shape, relative size, and distance between the two peaks of the bimodal probability density function of instantaneous velocities are not permanent and stable, but instead vary with the position of the velocity measurement.

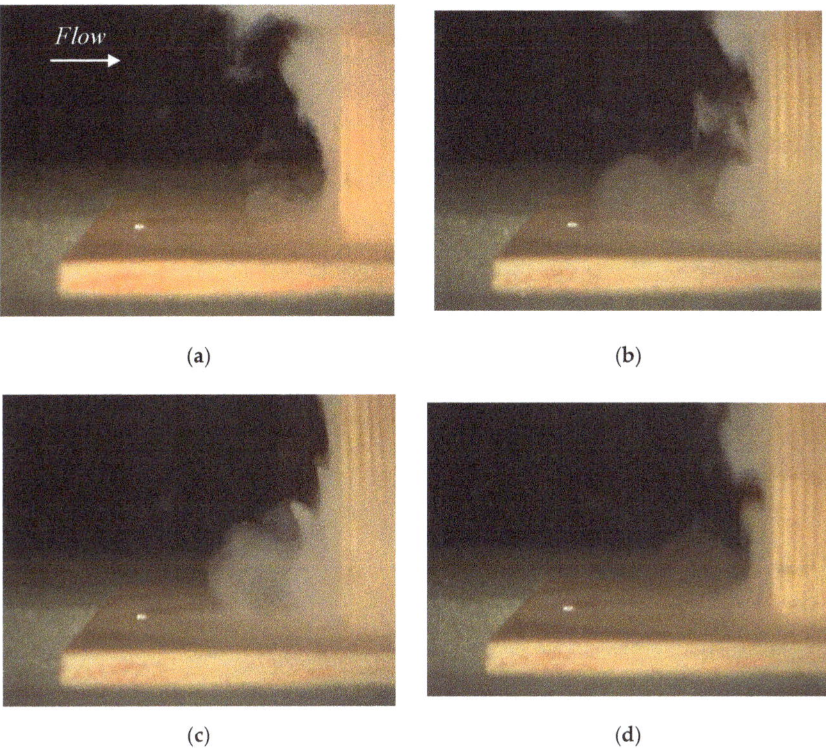

(a) (b)

(c) (d)

Figure 7. Flow visualization of horseshoe vortex with tracer injection in experimental Run 4: sequential snap shots from (**a–d**).

With respect to the turbulence characteristics just upstream of the pier in the horseshoe vortex region, quadrant analysis was also used to further characterize the turbulent events associated with the large-scale unsteadiness of the horseshoe vortex. Quadrant analysis was employed in this study by examining the joint probability density function of the streamwise and vertical components of fluctuating velocity, denoted as u' and w', respectively.

The four quadrants shown in Figure 8 correspond to four types of turbulent events which are defined as: I. outward interactions; II. ejections or bursts; III. inward interactions; and IV. sweeps that characterize the individual turbulent velocity measurements. The Reynolds stress is generally produced by all four types of events. The first quadrant event, I, is characterized by outward motion of high-speed fluid, with $u' > 0$ and $w' > 0$; the second quadrant event, II, is identified by outward motion of low-speed fluid, with $u' < 0$ and $w' > 0$, which is usually called ejection or bursts; the third quadrant event, III, is associated with inward motion of low-speed fluid, with $u' < 0$ and $w' < 0$; and finally, the fourth quadrant event, IV, represents the motion of high-speed fluid toward the bed, with $u' > 0$ and $w' < 0$, and it is called sweeps.

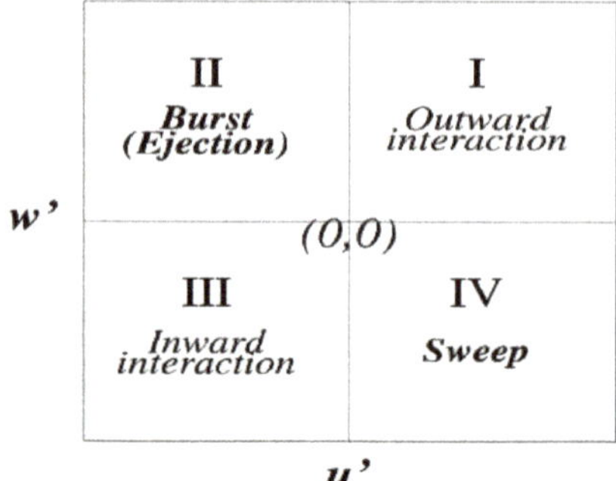

Figure 8. Schematic of the plate for quadrant analysis [36].

The relationships between the fluid motions and particle transfer near the wall can be elucidated through quadrant analysis. The ejections and sweeps contribute positively to the bed shear stress since $\overline{u'w'}$ is the flux of forward momentum to the bed, whereas the outward and inward interactions contribute negatively to the bed shear stress. The presence of a sweep corresponds to a local increase of the shear stress at the bed whereas the occurrence of an ejection corresponds to a local decrease of the shear stress at the bed. Therefore, the coherent sweep and ejection events appear to be responsible for transferring particles toward and away from the bed [37]. However, in nonuniform flows such as the wake region downstream of a backward-facing step, Keshavarzi et al. [36] have shown that turbulent events with the same shear stress contribute to different sediment transport rates due to the frequency structure of the turbulent events. While the horseshoe vortex is not expected to have the same turbulence structure as a turbulent wake, nevertheless, it does have the property of intermittency that is an important contributor to the scour process.

Among the velocity measurements made in experimental Run 2 and Run 4, two time series of velocity data at the nose of the pier were selected to conduct the quadrant analysis and to investigate the change of flow characteristics of the before-scour case and after-scour case. Figure 9 shows the results of Run 2. The time series analyzed in Figure 9a was measured for the fixed bed condition while the time series analyzed in Figure 9b was measured inside the scour hole under the same flow conditions at approximately the same distance above the bed. The joint probability density function in both figures are normalized to the peak values of 1.0, and it is necessary to multiply the contoured values by the scaling factors in the middle of the top of each figure to produce probability densities. The selected locations were both in the region of the horseshoe vortex and separated region upstream of the pier.

The results of quadrant analysis show a significant difference between the before-scour case and after-scour case in Figure 9a,b. The turbulent events for the before-scour condition were dominated by bursts and sweeps with a bimodal joint frequency distribution at the elevation of $z = 0.17\,b$. In Figure 9a before scour, sweeps events had a higher probability of occurrence, and for both types of events, the values of u' were greater than those of w' at the maximum probability of occurrence. Run 4 shows similar probability density patterns as in Figure 9. The bursts and sweeps are the primary forcing function for creating the scour hole because they both represent positive shear stress; however, the more important characteristic is the irregular oscillation between the two types of events as the horseshoe vortex alternately expands and contracts as the separation point moves back and forth, as shown in Figure 7. The sediment particles are lifted and entrained in an intermittent fashion as explained in Lee and Sturm [22]. In the equilibrium scour hole in Figure 9b, the turbulent events no longer display the bimodal distribution with all four types of events becoming approximately equally likely. The magnitude and frequency of all events are significantly affected by the change of the flow inside the scour hole. There is no effective event of the velocity fluctuations for altering the shear stress or moving the sediment out of the equilibrium scour hole. Although the existence of the bimodal distribution depends on the measuring location, it is significant that it disappears near the bed after scour and at the same distance above the bed as for the before-scour case.

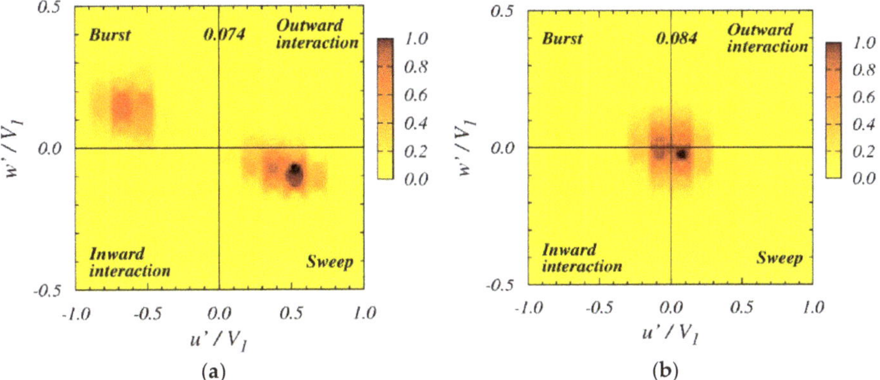

Figure 9. Joint probability density function of u' and w' for (**a**) before-scour case and (**b**) after-scour case measured at $x/b = -0.33$ and $z/b = 0.17$ for Run 2.

As a further indication of the differences in the turbulence properties as the scour hole develops, the integral time scale was computed for both the before-scour and after-scour time series associated with Figure 9. The integral time scale is defined as a measure of the time over which a velocity component is dependent on its past values and a rough measure of the time interval over which a fluctuating velocity component is highly correlated with itself; it is obtained by integration of the measured autocorrelation distribution over time and a measure of the memory of the process [38]. As given in Table 2, the integral time scale in the vertical direction for the before-scour case is considerably higher than for the after-scour case; whereas, in the streamwise direction, the time scale is the same order of magnitude for both cases. The vertical fluctuation in the before-scour case is a significant contributor to the processes of suspending and eroding sediment from the bed. Accordingly, a longer integral time scale of the velocity fluctuations gives more transport than a shorter integral time scale when the rate of sediment transport increases very rapidly at the initial stage of scour development.

Table 2. Comparison of integral time scales for before and after scour cases.

	Before Scour, Sec	After Scour, Sec
Streamwise direction	1.8	1.3
Vertical direction	4.3	0.7

4. Summary and Conclusions

Although local scour around bridge foundations have been extensively studied for several decades, there still remain problems because of difficulties in visualization and understanding the complex flow structure leading to scour. Thus, in this study, laboratory experiments were conducted with scaled-down bridge pier models for more complete descriptions of flow characteristics of the horseshoe vortex system around the complex bridge pier. During the experiments, velocities and turbulence intensities as well as the bed elevations before and after scour were measured by ADV. Furthermore, a simple visualization technique was used to capture the unsteadiness of horseshoe vortex. The results shows that horizontal velocity vector plot comparisons between before-scour and after-scour conditions measured at a certain relative height above the bed was not enough to explain the complex scour mechanism because the values were all temporally averaged ones, and so did not show the effects of the large-scale unsteadiness of the horseshoe vortex system upstream of a bridge pier. Further investigation using the probability distribution of instantaneous velocity components and quadrant analysis of velocity fluctuations in the horizontal and vertical directions shows that the vertical and streamwise velocity components exhibited a bimodal probability distribution before scour near the bed upstream of the pier where horseshoe vortex is responsible for sediment transport. The streamwise and vertical velocity fluctuations were observed to be dominated by sweeps and bursts, both of which contribute positively to the bed shear stress and exhibit a bimodal joint frequency distribution. The bursts and sweeps are the primary forcing function for creating the scour hole at initial stage because they both represent positive shear stress. They are the result of irregular oscillation of the horseshoe vortex between two preferred states as the reverse flow near the bed in front of the pier either extends upstream or retreats to a point closer to the pier. As a result, the sediment particles are lifted and entrained in an intermittent fashion. After scour is complete, at equilibrium stage, the turbulent events no longer display a bimodal distribution near the bottom of the scour hole. Furthermore, a quantitative physical connection is made between the scour depth and the large-scale unsteadiness of the horseshoe vortex system in front of the pier by comparing time scales for lifting of the sediment particle, and subsequent entrainment and transport of the particle out of the scour hole.

Even if this study provides additional insights to better understand the local pier scouring process and the relationship between scour depth and the horseshoe vortex, the local flow structures in the field are affected by additional parameters, such as vertical and lateral flow contraction, unsteadiness of discharge during the passage of a flood event, and their interaction. Furthermore, in reality, scouring often happens under live-bed scour conditions with infilling of the scour hole as the flood recedes which is not well understood over a long time series because it is greatly affected by the conditions required to generally mobilize the entire bed and is further complicated by the movement of bed forms through the bridge section. Thus, additional well-designed physical model and field measurements as well as numerical simulations are required to investigate these effects, including the modeling of flow contraction, realistic hydrographs, and bed forms, such as dunes and ripples.

Author Contributions: S.O.L. designed/fabricated the scaled-down bridge model, and conducted the flume experiments; S.O.L. and S.H.H. provided descriptions and motivation, and analyzed the laboratory data; S.H.H. wrote the paper; all authors participated in final review and editing of the paper.

Funding: This research received no external funding.

Acknowledgments: This work was partially supported by the National Research Foundation of Korea (NRF) grant funded by the Korea government (MIST) (NRF-2017R1A2B2011990). Also, Seung Ho Hong was supported by West Virginia University for this study.

Conflicts of Interest: The authors declare no conflict of interest.

References

1. Graf, W.H.; Istiarto, I. Flow pattern in the scour hole around a cylinder. *J. Hydraul. Res.* **2002**, *40*, 13–20. [CrossRef]
2. Chang, W.; Lai, J.; Yen, C. Evolution of scour depth at circular bridge piers. *J. Hydraul. Eng.* **2007**, *130*, 905–913. [CrossRef]
3. Gautam, P.; Eldho, T.I.; Mazumder, B.S.; Behera, M.R. Experimental study of flow and turbulence characteristics around simple and complex pier using PIV. *Exp. Therm. Fluid Sci.* **2019**, *100*, 193–206. [CrossRef]
4. Dey, S. Sediment pick-up for evolving scour near circular cylinders. *Appl. Math. Model.* **1996**, *20*, 534–539. [CrossRef]
5. Khaple, S.; Hanmaiahgari, P.; Gaudio, R.; Dey, S. Interference of an upstream pier on local scour at downstream piers. *Acta Geophys.* **2017**, *65*, 29–46. [CrossRef]
6. Unger, J.; Hager, W. Down-flow and horseshoe vortex characteristics of sediment embedded bridge piers. *Exp. Fluids* **2007**, *42*, 1–19. [CrossRef]
7. Dey, S.; Raikar, R.V. Characteristics of horseshoe vortex in developing scour holes at piers. *J. Hydraul. Eng.* **2007**, *133*, 399–413. [CrossRef]

8. Dargahi, B. Turbulent flow field around a circular cylinder. *Exp. Fluids* **1989**, *8*, 12. [CrossRef]

9. Simpson, R.L. Junction flows. *Annu. Rev. Fluid Mech.* **2001**, *33*, 415–443. [CrossRef]

10. Melville, B.W. Local Scour at Bridge Sites. Ph.D. Thesis, School of Engineering, University of Auckland, Aukland, New Zealand, 1975.

11. Baker, C.J. Theoretical approach to prediction of local scour around bridge piers. *J. Hydraul. Res.* **1980**, *18*, 12. [CrossRef]

12. Nakagawa, H.; Suzuki, K. Application of stochastic model of sediment motion to local scour around a bridge pier. In Proceedings of the 16th Congress of the International Association for Hydraulic Research, Sao Paulo, Brazil, 24 July–6 August 1975; pp. 285–299.

13. Einstein, H.A. *Bed-load Function for Sediment Transportation in Open Channel Flows*; Bulletin 1027 71, United States Department of Agriculture (USDA): Washington, DC, USA, 1950.

14. Qadar, A. Vortex Scour Mechanism at Bridge Piers. *Proc. Inst. Civ. Eng.* **1981**, *71*, 739–757. [CrossRef]

15. Kothyari, U.C.; Garde, R.J.; Ranga Raju, K.G. Temporal variation of scour around circular bridge piers. *J. Hydraul. Eng.* **1992**, *118*, 1091–1106. [CrossRef]

16. Ram, S. A theroretical model to predict local scour at bridge piers in non-cohesive soils. In Proceedings of the The Seventh International Symposium on River Sedimentation, Hong Kong, China, 16–18 December 1998; pp. 173–178.

17. Muzzammil, M.; Gangadhariah, T. Caracteristiques moyennes d'un vortex en fer a cheval au droit d'une pile cylindrique(The mean characteristics of horsehoe vortex at a cylindrical pier). *J. Hydraul. Res.* **2003**, *41*, 285–297. [CrossRef]

18. Kirkil, G.; Constantinescu, S.; Ettema, R. Coherent structures in the flow field around a circular cylinder with scour hole. *J. Hydraul. Eng.* **2008**, *134*, 572–587. [CrossRef]

19. Kirkil, G.; Constantinescu, G. Flow and turbulence structure around an instream rectangular cylinder with scour hole. *Water Res. Res.* **2010**, *46*, W11549. [CrossRef]

20. Schanderl, W.; Jenssen, U.; Strobl, C.; Manhart, M. The structure and budget of turbulent kinetic energy in front of a wall-mounted cylinder. *J. Fluid Mech.* **2017**, *827*, 285–321. [CrossRef]

21. Guan, D.; Chiew, Y.; Wei, M.; Hsieh, S. Characterization of horseshoe vortex in a developing scour hole at a cylindrical bridge pier. *Int. J. Sediment Res.* **2019**, *34*, 118–124. [CrossRef]

22. Lee, S.; Sturm, T.W. Effect of sediment size scaling on physical modeling of bridge pier scour. *J. Hydraul. Eng.* **2009**, *135*, 793–802. [CrossRef]

23. Hong, S.; Lee, S. Insight of Bridge Scour during Extreme Hydrologic Events by Laboratory Model Studies. *KSCE J. Civ. Eng.* **2018**, *22*, 2871–2879. [CrossRef]

24. Lee, S.; Hong, S. Reproducing field measurements using scaled-down hydraulic model studies in a laboratory. *Adv. Civ. Eng.* **2018**, *2018*, 11. [CrossRef]

25. Hong, S.; Abid, I. Scour around an erodible abutment with riprap apron over time. *J. Hydraul. Eng.* **2019**, *145*, 6. [CrossRef]

26. Hong, S.; Sturm, T.W.; Stoesser, T. Clear water abutment scour in a compound channel for extreme hydrologic events. *J. Hydraul. Eng.* **2015**, *141*, 12. [CrossRef]

27. Saha, R.; Lee, S.; Hong, S. A comprehensive method of calculating maximum bridge scour depth. *Water* **2018**, *10*, 1572. [CrossRef]

28. Yoon, K.; Lee, S.; Hong, S. Time-averaged turbulent velocity flow field through the various bridge contractions during large flooding. *Water* **2019**, *11*, 13. [CrossRef]

29. Chua, K.; Fraga, B.; Stoesser, T.; Hong, S.; Sturm, T.W. Effect of bridge abutment length on turbulence structure and flow through the opening. *J. Hydraul. Eng.* **2019**, *145*, 19.

30. Ge, L.; Lee, S.; Sotiropoulos, F.; Sturm, T.W. 3D unsteady RANS modeling of complex hydraulic engineering flows. Part II: Model validation and flow physics. *J. Hydraul. Eng.* **2005**, *131*, 809–820. [CrossRef]

31. Jones, J.S.; Kilgore, R.T.; Mistichelli, M.P. Effects of footing location on bridge pier scour. *J. Hydraul. Eng.* **1992**, *118*, 280–290. [CrossRef]

32. Melville, B.W.; Raudkivi, A.J. Effects of foundation geometry on bridge pier scour. *J. Hydraul. Eng.* **1996**, *122*, 203–209. [CrossRef]

33. Coleman, S.E. Clearwater local scour at complex piers. *J. Hydraul. Eng.* **2005**, *131*, 330–334. [CrossRef]

34. Devenport, W.J.; Simpson, R.L. Time-dependent and time-averaged turbulence structure near the nose of a wing-body junction. *J. Fluid Mech.* **1990**, *210*, 23–55. [CrossRef]

35. Paik, J.; Escauriaza, C.; Sotiropoulos, F. On the bimodal dynamics of the turbulent horseshoe vortex system in a wing-body junction. *Phys. Fluids* **2007**, *19*, 045107. [CrossRef]
36. Keshavarzi, A.; Melville, B.; Ball, J. Three-dimensional analysis of coherent turbulent flow structure around a single circular bridge pier. *Environ. Fluid Mech.* **2014**, *14*, 821–847. [CrossRef]
37. Marchioli, C.; Soldati, A. Mechanisms for particle transfer and segregation in a turbulent boundary layer. *J. Fluid Mech.* **2002**, *468*, 283–315. [CrossRef]
38. Kundu, P.K. *Fluid Mechanics*; Academic Press: San Diego, CA, USA, 1990.

Article

Turbulent Flow Field around Horizontal Cylinders with Scour Hole

Nadia Penna *, Francesco Coscarella and Roberto Gaudio

Dipartimento di Ingegneria Civile, Università della Calabria, Via Pietro Bucci, 87036 Arcavacata, Rende CS, Italy; francesco.coscarella@unical.it (F.C.); gaudio@unical.it (R.G.)
* Correspondence: nadia.penna@unical.it; Tel.: +39-0984-496-553

Received: 21 November 2019; Accepted: 31 December 2019; Published: 2 January 2020

Abstract: This study presents the results of an experimental investigation on the flow-structure interactions at scoured horizontal cylinders, varying the gap between the cylinder and the bed surface. A 2D Particle Image Velocimetry (PIV) system was used to measure the flow field in a vertical plane at the end of the scouring process. Instantaneous and ensemble-averaged velocity and vorticity fields, viscous and Reynolds stresses, and ensemble-averaged turbulence indicators were calculated. Longitudinal bed profiles were measured at the equilibrium. The results revealed that suspended and laid on cylinders behave differently from half-buried cylinders if subjected to the same hydraulic conditions. In the latter case, vortex shedding downstream of the cylinder is suppressed by the presence of the bed surface that causes an asymmetry in the development of the vortices. This implies that strong turbulent mixing processes occur downstream of the uncovered cylinders, whereas in the case of half-buried cylinders they are confined within the scour hole.

Keywords: horizontal cylinder; turbulence structures; scour

---

1. Introduction

Although flow around a cylinder is one of the classical subjects of fluid dynamics, few investigations have focused on the analysis of the turbulence structures of a steady flow at scoured horizontal cylinders. This topic is relevant in the hydraulic field because it is encountered in many engineering applications, such as pipelines suspended, laid on or half-buried installed across mobile riverbeds, where a small depth-to-diameter ratio is most relevant and scouring occurs under unidirectional current [1]. In all these cases, the 3D flow field is extremely complicated due to the separation and the creation of multiple vortices. However, the complexity is further exaggerated owing to the dynamic interaction between the flow and the movable bed [2]. Erosion may occur around the pipelines, causing a higher gap between it and the bed surface and, therefore, compromising their safety. Several research findings can be found in literatures for estimating the local scour around underwater pipelines under unidirectional current [3]. Some of these studies have shown that the scour depth under unidirectional current is always higher than that under pure wave action or the combined effect of wave and current with the same bottom shear stress [3]. Accurate estimates of the scour depth are important because flow-induced oscillation by wake-vortex shedding may provoke fatigue failure of the pipeline itself [4], which is subjected to additional unsteady forces such as lift and drag.

The first investigation for estimating scour depth at submarine pipeline was conducted by Chao and Hennessy [5], who proposed an analytical method for estimating the maximum scour depth under pure current condition. The use of this method was supported by other authors e.g., [6–8]. The main drawback of the method is the use of a potential flow theory in deriving the solution. In real flow, the fluid is not inviscid, and separation occurs at the rear of the pipe. This phenomenon affects the flow conditions [3]. Later, Kjeldsen et al. [9] proposed an equation that implies that the scour depth only depends on flow velocity and pipe diameter, but excludes the effect of flow depth and grain size. Bijker and Leeuwestein [10] stated that the scour depth depends on the undisturbed flow velocity, pipe diameter, flow depth, height of pipe above bed level, and grain size. They stated that the principal cause of erosion is a local increase in transport capacity of the water passing a pipeline, while deposition occurs where this capacity decreases [3]. Ibrahim and Nalluri [11] proposed two empirical equations relating the scour depths to flow parameters for both clear-water and live-bed conditions. The equations were derived purely from curve fitting technique. Furthermore, they stated that the grain size has no influence on the scour depth, apart from the indirect influence on the value of the critical velocity. Hansen et al. [12] obtained a relationship between the bed velocity in the scour hole and the two geometric quantities e/D and d_s/D, where e is the gap between the underside of the cylinder and the original bed level, D is the cylinder diameter and d_s the scour depth. They stated that the flume width would also influence this relationship. Mao [13] examined the scour profiles below pipelines under different flow velocities and observed that, for $d_s/D < 1$ d_s, is a weak function of the flow Shields parameter. He also identified two cases of the scour process: (1) jet period, which decides the maximum scour depth; (2) wake period, which decides the location of maximum scour depth. Maza [14] proposed a graphical solution for the estimation of d_s that is a function of the initial gap-pipe diameter ratio and the flow Froude number, while Moncada and Aguirre [15] gave an empirical equation of d_s assuming a similar functional representation used by Maza [14]. Chiew [16] identified that piping plays a dominant role in initiating scour at submarine pipelines. Later, Chiew [3] proposed an empirical function relating the flow depth ratio, h/D, with the gap-flow rate ratio, which can be used to determine the amount of gap flow through the scour hole at equilibrium conditions. Li and Cheng [17] used the finite difference method to solve the Laplace equation of velocity potential and a boundary adjustment technique to calculate the scour profiles below pipelines, while Brørs [18] used the finite element method to simulate the scour profiles below pipelines. Dey and Singh [4] examined their experimental results to describe the influence of various parameters on the equilibrium scour depth, that is flow depth, sediment gradation, different shaped cross-sections of pipes. Lately, Mohr et al. [19] related the rate of scour beneath a pipeline to the fundamental erosion properties of the sediment, namely the transport rate along the bed and the true erosion rate of the sediment. These arguments lead to two new empirical formulas that may be used to predict the time scale of the scour process beneath subsea pipelines. Note that these previous researches concentrate on soil scour around fixed pipelines. More recently, Gao et al. [20] simulated experimentally the current-induced sand scour around a vibrating pipeline to further investigate the mechanism of the coupling effects between pipe vibration and sand scour.

At the same time, several studies have been carried out both numerically and experimentally on the flow field analysis considering a flat bed. A state-of-art review of the research on the cylinder-bed surface interactions exposed to currents was conducted by Fredsøe [1]. For example, Bearman and Zdravkovich [21] carried out an experimental study on the flow around a cylinder lying horizontally at various elevation above a plane bed surface. They demonstrated that regular vortex shedding was suppressed for all gaps less than about $0.3D$. For gaps greater than $0.3D$, the Strouhal number, that is the ratio between inertial forces due to the unsteadiness of the flow and expresses the oscillating flow mechanisms, was found to be remarkably constant and the only influence of the plate on vortex shedding was to make it a more highly tuned process as the gap was reduced. Later Zdravkovich [22] studied in detail the flow separation from a flat plate induced by a circular cylinder. He demonstrated that when the cylinder was placed above the bed surface, the downstream separation region contained two counter-rotating vortices separated by stagnant fluid. However, no regular vortex shedding was observed in this configuration, whereas he reported that in the case of turbulent boundary layer vortex shedding occurred for a gap of $0.2D$ and for the laminar boundary layer at a gap of $0.3D$. Lei et al. [23] analyzed the hydrodynamic forces and vortex shedding of a cylinder at different locations in the boundary layer. They proposed a quantitative method for identifying the vortex shedding suppression point. Their observations showed that the vortex shedding is suppressed at a gap of about 0.2–0.3D, depending on the thickness of the boundary layer. This critical gap decreases as the thickness of the boundary layer increases. Price et al. [24] performed visualization studies on the flow-cylinder interactions in the boundary layer. Specifically, they distinguished four distinct regions: (i) for gaps < 0.125D, the gap flow is suppressed or extremely weak, and separation of the boundary layer occurs both upstream and downstream of the cylinder. Although there is no regular vortex shedding, there is a periodicity associated with the outer shear-layer; (ii) in the region between 0.125D and 0.5D, the flow is very similar to that for very small gaps, except that there is a pronounced pairing between the inner shear-layer shed from the cylinder and the wall boundary layer; (iii) the region between 0.5D and 0.75D is characterized by the onset of vortex shedding from the cylinder; (iv) for the fourth region, with a gap greater than D, there is no separation of the wall boundary layer, either upstream to or downstream of the cylinder. However, downstream of the obstacle, alternate vortex shedding from the cylinder affects the wall boundary layer. Hatipoglu and Avci [25] studied experimentally and numerically the flow around a horizontal cylinder mounted and partially buried, showing that the lengths of the separation regions near the upstream and downstream of the cylinder decreased with the increasing burial ratio. More recently, Akoz et al. [26] studied quantitatively the flow characteristics of the circular cylinder laid on a fixed surface, by using the Particle Image Velocimetry (PIV) technique. The main purpose was to reveal the mechanisms of vortical flow structures that are mostly responsible for scour and burial processes. They demonstrated that the intersection of the bed surface and cylinder enhances the burial mechanisms hydrodynamically even in wake flow regions. Furthermore, it was shown that the wake flow region is shortened in size in the longitudinal direction as a function of the Reynolds number. More recently, Arslan et al. [27] studied for four different submergence ratios the 3D unsteady flow around a rectangular cylinder with the large eddy simulation (LES) turbulence model. To conduct the analysis, they used the experimental data obtained by Malavasi and Guadagnini [28]. As a result, they characterized the behavior of vortex structures generated by separated flow and the hydrodynamic forces acting on the semi-submerged structure.

Few are the studies in which the turbulence flow structures are analyzed focusing in the neighborhood of the cylinder considering a mobile bed. The first studies on this topic were those of Kjeldsen et al. [9] and Mao [13], who examined the flow around a pipe placed over a scour hole, Jensen et al. [29] investigated experimentally the flow around a pipeline placed initially on a flat bed at five characteristic stages of the scour process in currents. They found that the mean flow field and turbulence around and the force on a pipeline undergo considerable changes, as the scour below the pipeline develops in time and space. Moreover, vortex shedding comes into existence at early stages of scour process, first in a somewhat premature form caused by the close proximity of the dune formed downstream the pipe as a result of deposition of scoured material. As the dune moves away from the pipe, the vortex shedding gradually reaches a stage which resembles the shedding process of a free cylinder. Starting from this experimental campaign, Smith and Foster [30] examined the flow physics around the pipeline (considering also the bed scouring phenomenon) using numerical modelling. They compared the mean horizontal and vertical velocities and the wake characteristics measured by Jensen et al. [29] to those obtained by applying a computational fluid dynamics (CFD) model (Flow3D) using both a two-equation k-ε model and a Smagorinsky LES turbulence closure scheme for five stages of the scour process.

Thus, the main purpose of the study was to identify the flow structures in shallow water condition developed once the bed scour reached the equilibrium, varying the gap between the cylinder and the bed surface. Specifically, three different conditions were reproduced in a laboratory flume: a suspended cylinder, a laid on cylinder and a half-buried cylinder. The analysis was focused on the velocity measurements in correspondence of the horizontal cylinder and on the determination of the instantaneous and ensemble-averaged velocity fields, the instantaneous and ensemble-averaged vorticity fields, viscous and Reynolds stresses, and the ensemble-averaged turbulence indicators. In addition, the equilibrium longitudinal bed profiles were discussed in order to better comprehend the effect induced by the flow-structure interactions.

2. Experimental Set-Up and Procedure

The experimental campaign was performed at the Laboratorio "Grandi Modelli Idraulici" (GMI, Università della Calabria, Italy). A recirculating flow channel (9.6 m long, 0.485 m wide, 0.5 m deep) was used in this study. The flume side walls were made of glass in order to visualize the flow. The inlet of the flume comprised of a stilling tank, an uphill slipway, and honeycombs (having a diameter of 10 mm) to dampen the flow disturbances at the entry. The test section was located at 7.33 m downstream of the flume entrance and it was 0.165 m long. The flow depth was regulated by a downstream tailgate. To collect the outflow, a tank equipped with a calibrated Thomson weir to measure the flow discharge was attached downstream of the tailgate.

The bed was constituted by very coarse sand having a median sediment size d_{50} of 1.53 mm (0.06 mm $< d <$ 2 mm, where d is the size of sediments) and geometric standard deviation σ_g [$=(d_{84}/d_{16})^{0.5}$] of 1.24, where d_{16} and d_{84} are the 16% and 84% (by weight) finer sizes of sediments, respectively. The sediment density was ρ_s = 2680 kg/m^3. To prepare the bed, sediments were initially spread within the flume and screeded to create a bed with a longitudinal slope S_0 of 0.1%.

A horizontal cylinder of 30 mm in diameter made of Plexiglas was set up at the center of the test section. The cylinder was fixed to the flume walls and kept normal to the flow direction. Three different gaps, respectively e_1, e_2 and e_3, between the cylinder and the bed surface were considered, as represented in Figure 1 and reported in Table 1: (i) suspended cylinder; (ii) cylinder laid on the bed; (iii) partially buried cylinder. To ensure direct comparisons of the results, the bed surface was leveled at the start of each experiment.

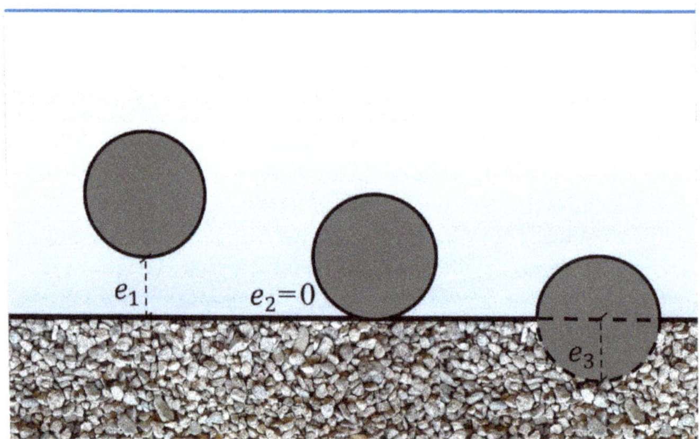

Figure 1. Schematic representation of the geometrical conditions of the experimental runs, where e_i (i = 1, 2, 3) is the gap between the lower edge of the cylinder and the bed surface.

Table 1. Geometric and hydraulic parameters of the experimental runs.

Parameter (Units)	Run 1	Run 2	Run 3
e_i (m)	0.015	0	−0.015
h (m)	0.073	0.073	0.076
Q (m^3/s)	0.010	0.010	0.010
U (m/s)	0.28	0.28	0.27
U_c (m/s)	0.37	0.37	0.37
u_* (m/s)	0.026	0.026	0.025
Re	7.9×10^4	7.9×10^4	7.9×10^4
Re_c	8.1×10^3	8.1×10^3	7.8×10^3
Re_*	85	85	82
Fr	0.33	0.33	0.31

The runs were carried out under the same hydraulic conditions. Specifically, the initial flow depth h was about 0.074 m, as measured immediately upstream to the cylinder from the sediment crest. Therefore, the aspect ratio B/h was greater than 5 (B being the flume width) and the effect of the sidewalls is negligible [31] in our experimental set-up. However, the ratio h/D was kept constant and equal to 2.5. Thus, being $h/D < 5$ [4], the shallowness effect is present in all the runs and, therefore, the resulting scouring process is due to the cylinder-bed surface interactions at different gaps in shallow water condition. The threshold flow velocity, U_c, was determined using Neill's empirical formula [32] and was equal to 0.37 m/s, higher than the average flow velocity $U = Q/(Bh) \approx 0.28$ m/s, Q being the flow discharge. This indicates a clear-water condition. The flow Reynolds number Re (=$4Uh/v$, where $v = 1.04 \times 10^{-6}$ m^2/s at 18.4 °C is the kinematic water viscosity), the flow Froude number Fr [=$U/(gh)^{0.5}$, where g is the gravitational acceleration], the cylinder Reynolds number Re_c (=UD/v, where D is the cylinder diameter) and the other hydraulic parameters are listed in Table 1 for each experimental run. Specifically, upstream of the cylinder (where the flow was not affected), the shear velocity u_* was estimated by extending the Reynolds shear stress profile linearly to the maximum gravel crest as $\left(-\overline{u'w'}\right)^{0.5}\Big|_{z=z_c}$ (where u' and w' are the temporal velocity fluctuations in the streamwise and vertical directions, respectively, and z_c is the maximum gravel crest elevation). Hence, the Reynolds number of the sediments Re_* was calculated as $u_*\varepsilon/v$, where ε is the equivalent sand roughness height $\approx 2d_{50}$ [33].

During all the experimental runs bed scouring occurred. It was found that the equilibrium scour depth was reached within 24 h for all the three experimental runs. This is in line with the findings of Chiew [3]. Dey and Singh [4] demonstrated that the time to reach the equilibrium scour below a pipeline is shorter than that at an abutment, since it is due to strong pressurized flow and not to primary or horseshoe vortex. Also, Mao [13] observed equilibrium scour below a cylinder in just over 3 h.

A 2D PIV system manufactured by TSI was used to measure the flow field. It consisted of a 12 bit CCD camera (Nikon, city, country, 50 mm F1.8 lens, 2048 × 2048 pixel^2 resolution, 15.2 × 15.2 mm^2 sensor size) and a double pulse Nd:YAG laser with a frame rate of 15 Hz and a pulse energy of 200 mJ at a wavelength of 532 mm. Titanium dioxide powder, having a diameter of 3 μm and a mass density of 4.26 kg/m^3, was used as a tracer during flow measurements. The test section was illuminated with the laser to capture the movement of the tracer particles through the camera placed parallel to the laser sheet. The flow measurements were taken in correspondence of the horizontal cylinder. In order to measure the flow field, 3000 pairs of images were captured over a period of about 200 s. The inter-frame time between two laser pulses was set equal to 1400 μs, which means that, considering $U \approx 0.28$ m/s, we were able to measure only eddies greater than 0.39 mm. The actual size of the eddy is imposed by the spatial resolution of the PIV measurements [34]. The field of view was 165 × 165 mm^2 with an interrogation area of 32 × 32 pixel^2. A 50% overlap of the interrogation areas was employed to increase the spatial resolution of the measurement [35] to about 1.3 mm (which corresponds to 12.4 pixel/mm). The Insight 4G-2DTR software was used during the acquisition phase and to process the resulting data.

The precision in the flow velocity measurements with a PIV system depends to a great extent on the errors introduced by the sub-pixel estimator in the cross-correlation. This is known as "peak-locking" and is a bias error in the PIV that occurs when the particle images are too small, causing the particle image displacement to be biased toward integer value. The error is estimated to be 10% of the particle image diameter, which is the diameter in pixels of the particle as seen through the camera [36]. The mean particle image diameter in the present case is about 1 pixel, and a typical displacement between the cross-correlation image pairs is 5 pixels in the main flow direction. Therefore, the estimated random error of the measured velocity vector in each interrogation area is about 2% for the streamwise velocity component.

However, as *a priori* method, in order to reduce peak-locking, through a trial and error procedure we set an aperture number $f\#$ equal to 11, in order to ensure that all the tracer particles in the light sheet (having a thickness of 2 mm) were in focus. Nevertheless, following the procedure described in Padhi et al. [34], we used also and *a posteriori* method pre-processing all the images by using a filter (already available in the Insight 4G-2DTR software), which optimized the particle image diameter with respect to the peak estimator, in order to improve the signal-to-noise ratio.

The flow was stopped at the end of the measurement phase and, subsequently, the flume was slowly emptied without any disturbance for the bed scour by using a bottom outlet located 1 m downstream of the test section. For each run, the bed surface was acquired with the photogrammetry technique. Photogrammetry is used in different fields such as topographic mapping, architecture, engineering, cultural heritage and geology e.g., [37–39]. Following the procedure described in Penna et al. [40], a Nikon D5200 camera was used, equipped with a Nikkor 18–55 mm f/3.5–5.6 G VR lens. The resulting 3D point cloud was at first transformed into an unstructured triangular mesh by using the software PhotoScan (Agisoft, St. Petersburg, Russia). Then a structured grid was extracted using the commercial software Rhinoceros (McNeel & Associates, Seattle, WA, USA), with a spatial resolution $\delta l = 5$ mm in both the streamwise and spanwise directions.

3. Results and Discussion

3.1. Longitudinal Bed Profiles

The centerline longitudinal bed profiles at equilibrium, for each experimental run, are shown in Figure 2. Here, the origin of the x-axis (streamwise direction) is set in correspondence of the cylinder center, whereas the z-axis indicates the vertical direction. Specifically, the horizontal and vertical axes were made dimensionless by dividing them by the cylinder diameter. Furthermore, $z/D = 0$ is the bed surface at the start of each experiment. These longitudinal profiles were extracted from the bed surface models derived from the photogrammetry technique.

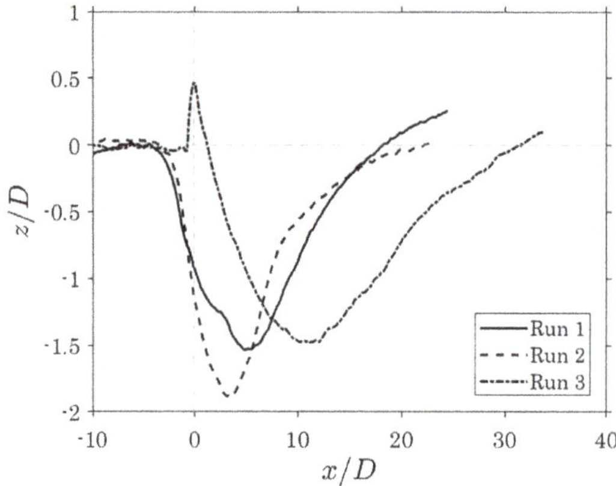

Figure 2. Centerline longitudinal bed profiles at equilibrium.

The data clearly highlight how the gap between the cylinder and the bed surface influences the scour formation and, thus, the bed profile. A steady flow that impacts on a cylinder laid on the bed causes a deeper scour hole than in the case in which the cylinder is suspended. Specifically, it was $1.5D = 4.6$ cm and $1.9D = 5.7$ cm for Run 1 and Run 2, respectively. This means that the gap in Run 1 allows the flow to pass with less disturbance to the sand bed. It is also evident that the maximum equilibrium scour depth d_s occurs closer to the cylinder in the second case than in Run 1. The distance between the center of the cylinder and the maximum equilibrium scour depth, x_s, is $5D = 15$ cm for Run 1 and $3.5D = 10.5$ cm for Run 2. The behavior can be explained by analyzing the vortex system induced by the fluid-structure interaction, which is described in the following sections. Furthermore, it is possible to note that the resulting scour hole involves the area upstream to and beneath the cylinder, owing to a particular process also known as tunnel erosion [41], that is ascribed to seepage processes [2,42] and occurs when the ratio between the flow depth and the cylinder diameter is less than 3.5 [16]. The vortices due to the obstacle led to instabilities to the particles, moving them away. The consequent erosion process is due to the increased velocity underneath the cylinder.

A different behavior can be noted considering Run 3 in which the cylinder is half-buried: tunnel erosion does not occur and the scour hole develops downstream of the cylinder itself. Hence, in this case, the shape of the bed profile is comparable to that formed in the presence of a bed sill subjected to a steady flow. This will be further explained analyzing the flow field around the cylinder in the following sections. Interestingly, even if d_s is the smallest value ($d_s = 1.5D = 4.4$ cm) if compared to the other two cases, the scour length l_s is about $0.3D = 0.9$ m and, therefore, the scouring process affects a larger amount of sediments than in Run 1 and Run 2, for which it is almost $0.22D = 0.65$ m. Note that the small hump visible in Figure 2 at $x/D = 0$ is the profile of the cylinder itself.

The obtained bed profile at equilibrium for $e = 0$ (Run 2) was compared with the experimental data of Chiew [3], Gao et al. [20] and Dey and Singh [4]. To better understand the different behavior that can be observed in Figure 3, Table 2 shows the experimental conditions adopted for each run. The bed profiles of Gao et al. [20] and Dey and Singh [4] were determined in deep-water condition ($h/D > 5$), whereas those related to the studies conducted by Chiew [3] and of the present work were in shallow-water condition ($h/D < 5$). Therefore, it is easy to note that for shallow-water condition a deeper and longer scour hole is formed, leaving aside for the moment considerations about the other parameters involved. Furthermore, in the case of deep-water condition the scour hole develops to great extent below the cylinder, whereas for shallow-water the maximum scour depth is reached downstream of it. By comparing the bed profile of Run 2 and that of Gao et al. [20], which are characterized by similar cylinder diameter and flow velocity, the higher water depth of Gao et al. [20] produced a maximum scour depth of about 1.8 cm, whereas that obtained in the present study was about 5.7 cm, although d_{50} was 4 times less than 1.53 mm of Run 2. However, it must be pointed out that also the considered equilibrium time (t_e) can influence the observed discrepancies between the bed profiles. Based on results obtained from Chiew [3], only 50–70% of the equilibrium scour depth is reached in three to four hours of testing. Finally, by comparing the bed profile of Run 2 and that of Dey and Singh [4], which are characterized by a similar sediment size, it is possible to note that the maximum scour depth is of the same order, despite the fact that D and U are higher than those of the present study. This can again be ascribed to both the water depth and also to the equilibrium time.

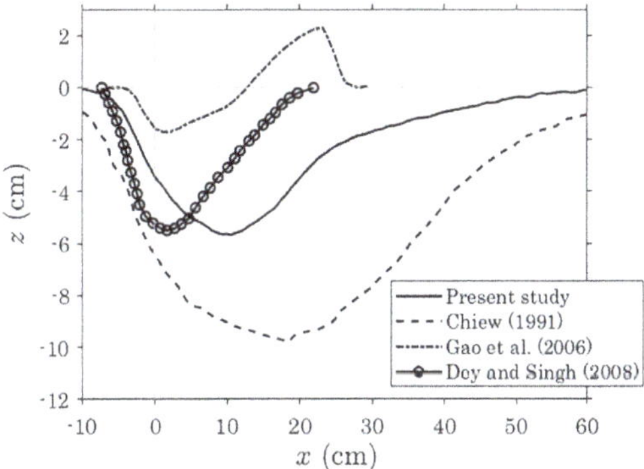

Figure 3. Comparison between centerline longitudinal scour profiles around a horizontal cylinder with $e = 0$.

Table 2. Geometric and hydraulic parameters of the literature experimental runs for $e = 0$.

	D (cm)	d_{50} (mm)	h (cm)	U (m/s)	t_e (h)
Present study	3	1.53	7.3	0.280	24
Chiew (1991) [3]	3.2 ÷ 6.3	0.33 ÷ 1.70	5 ÷ 18	-	>24
Gao et al. (2006) [20]	3.2	0.38	40	0.255	4
Dey and Singh (2008) [4]	4	1.86	20	0.497	12

3.2. Instantaneous and Ensemble-Averaged Velocity Fields

A sequence of four consecutive instantaneous flow fields at different times t for all the experimental runs are shown in Figure 4. The velocity has a magnitude $|\mathbf{u}| = \left(u^2 + w^2\right)^{0.5}$, where u and w are the instantaneous streamwise and vertical velocity components, respectively. The velocity was made dimensionless by dividing it by U. The origin of the x-axis refers to the start of the test section acquired with the PIV system. Both the axes were made dimensionless by dividing by D. Furthermore, Figure 4 shows the streamlines for each velocity field. Note that invalid vectors were removed from the instantaneous velocity fields (blank areas in the colormaps), as well as those below and along the edge of the cylinder since they corresponded to an area poorly illuminated by the laser. Vectors identified as spurious were replaced by the ensemble-averaged values when computing the average over time.

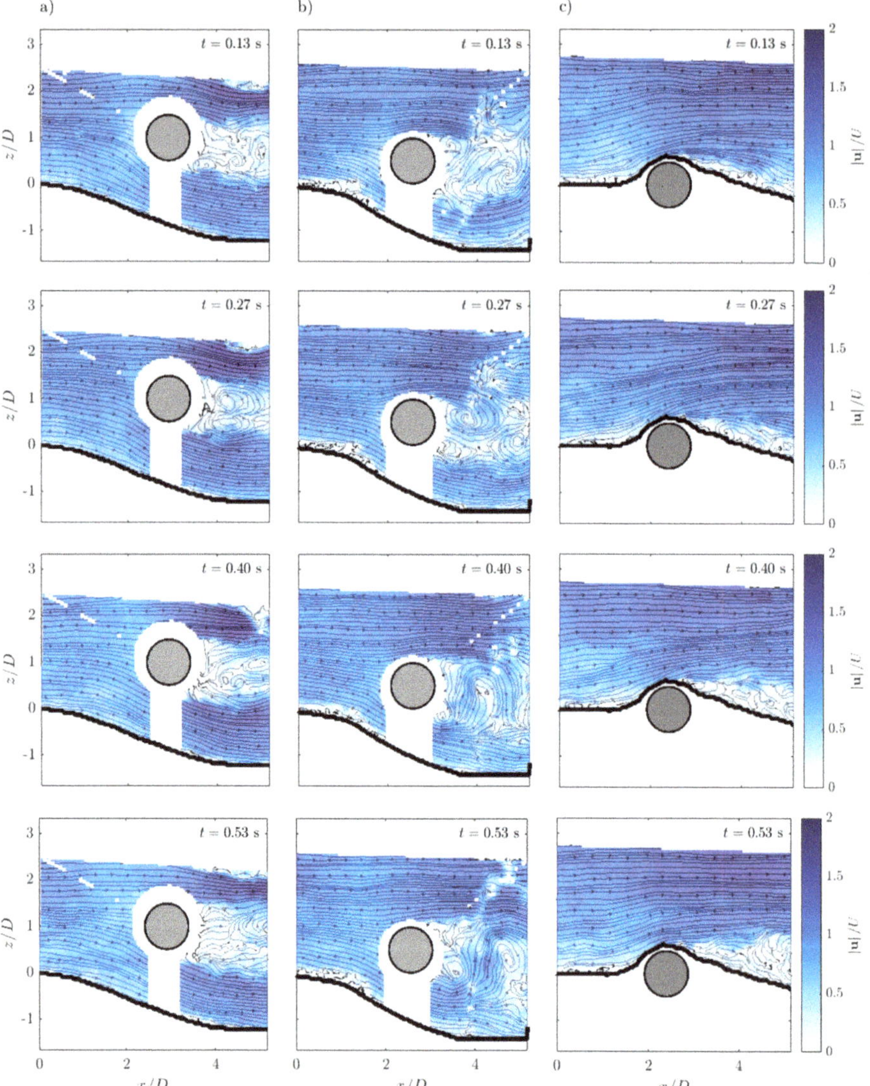

Figure 4. Dimensionless instantaneous velocity fields at different times t on a vertical plane in (**a**) Run 1, (**b**) Run 2 and (**c**) Run 3.

From the analysis of Figure 4, it is evident that upstream to the cylinder, at the beginning of the study area, the flow velocities, in all the three runs, increase with the vertical distance z, as usually observed in an open-channel flow, and the streamlines are nearly horizontal. Thus, in both Run 1 and Run 2, the incoming flow approaches the upstream face of the cylinder and causes flow separation. The main flow is divided into two flows: one part is oriented towards the bed surface (this is the main cause of scouring) and the other part is oriented towards the free surface [26], giving birth to two separation zones. Since $300 < Re_c < 3 \times 10^5$ (Table 1), the boundary layer over the cylinder surface is laminar. For Run 2 the separation zone at the free-stream side of the cylinder moves upstream with respect to that of Run 1. This is in agreement with previous findings [26,29,43]. A different behavior occurs for Run 3: the obstruction represented by the cylinder creates an adverse pressure gradient resulting in the separation of flow lines forming a vortex on the upstream front of the cylinder. Separation of flow also takes place at the free-stream side of the cylinder. Reverting to Run 1 and Run 2, a pair of vortices in the wake of the cylinder is noticeable, causing vortex shedding. The vortices are shed alternately at both sides of the cylinder at a certain frequency. Specifically, following the classification of Sumer and Fredsøe [43], the generated wake is completely turbulent, whereas the vortex-shedding frequency, f, can be derived from the Strouhal number:

$$St = \frac{f \times D}{U} \tag{1}$$

The Strouhal number in proximity of a wall may change with respect to the case in which no wall is present. Shedding frequency tends to increase (yet slightly) as the gap ratio decreases [44]. However, the flow field in this study was analyzed at the equilibrium phase, that is when the scouring phenomena was already exhausted. This means that the gap between the cylinder and the bed is 4.27 cm and 3.43 cm for Run 1 and Run 2, respectively (Figure 2). Therefore, since the gap ratio e_i/D is higher than 1 for Run 1 and Run 2 (1.42 and 1.14, respectively), it is possible to assume that in these cases the Strouhal number in proximity of the wall is equal to the Strouhal number for wall-free cylinder [44,45]. Note that the vortex shedding occurs in a region where the scour hole is deeper than that at the cylinder itself ($e_1/D > 1.42$ and $e_2/D > 1.14$ for Run 1 and Run 2, respectively). Therefore, the wall represented by the mobile bed has no effects on the Strouhal number, even if concave and not flat. Considering that for a smooth circular cylinder St remains practically constant at the value of 0.2 for $300 < Re_c < 3 \times 10^5$ [46], the vortex-shedding frequency can be determined as $St \times U/D$, thus 1.87 s^{-1} for Run 1 and Run 2. This means that vortex shedding occurs about twice per second. Figure 4 helps in visualizing the formation and destruction of the vortex system, from $t = 0.13$ s to 0.53 s, with $t = 0$ at the beginning of the experimental measurements. The vortex originated from the free-stream side of the cylinder (e.g., Figure 4b at $t = 0.13$ s) is strong enough to draw the opposing vortex (at the cylinder wall-side) downstream across the wake region (e.g., Figure 4b at $t = 0.27$ s). Note that the streamlines coming from above are oriented in the clockwise direction, while the others in the anti-clockwise direction. The opposite sign will then cut off further supply of vorticity to the upper vortex from its boundary layer [43], and therefore this vortex is shed and the transported downstream by the flow (e.g., Figure 4b at $t = 0.40$ s). Thus, a new vortex forms at the free-stream side of the cylinder. However, the lower vortex is now stronger than this new upper vortex and, therefore, this will lead to its shedding (e.g., Figure 4b at $t = 0.53$ s). Instead, for Run 3 a reverse roller is formed in the wake region. Vortex shedding is suppressed by the presence of the bed surface that causes an asymmetry in the development of the vortices [43].

To analyze the ensemble-averaged flow field in a spatial flow domain, Figure 5 shows the streamlines and the contours of the velocity having a magnitude $|\overline{\mathbf{u}}| = \left(\overline{u}^2 + \overline{w}^2\right)^{0.5}$, where $\overline{u}$ and $\overline{w}$ are the ensemble-averaged streamwise and vertical velocity components, respectively. The ensemble-averaged velocity was made dimensionless by dividing it by U.

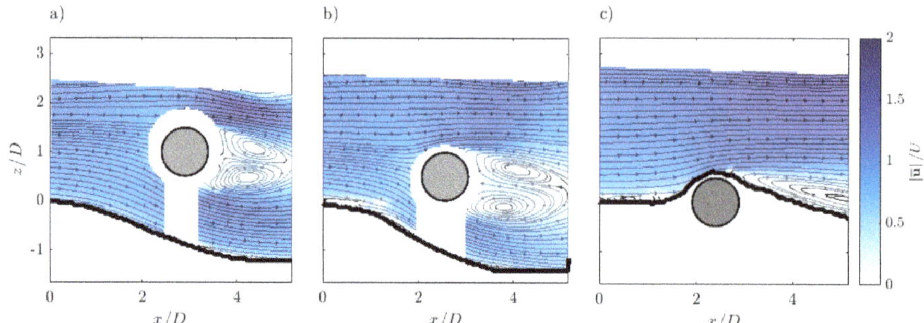

Figure 5. Dimensionless ensemble-averaged velocity fields on a vertical plane in (**a**) Run 1, (**b**) Run 2 and (**c**) Run 3.

The analysis of Figure 5 recalls the separation of the main flow upstream to the cylinder in both Run 1 and Run 2, described on the basis of the instantaneous velocity fields of Figure 4. It is apparent that the increase of the gap between the cylinder and the bed surface (Run 1) causes the formation of two opposite vortices in the wake of the cylinder shorter and more compressed than those of Run 2. This behavior may be attributed to the fact that in Run 1 the cylinder is close to the water surface, pushing down the upper vortex. In the meantime, a greater amount of flow passes below the cylinder and, therefore, tends to expand immediately downstream of it, compressing the vortex system in the wake region. The velocity at the underside of the cylinder is slightly lower than the one at the top side. This is in agreement with the observations of Jensen et al. [29]. The ensemble-averaged velocity field of Run 3, instead, shows the roller vortices upstream to and downstream of the cylinder in the scour hole, as discussed before. This led to a flow acceleration along the water depth above the recirculation zone.

Figure 6 shows the vertical profiles of the ensemble-averaged streamwise flow velocity at different distances upstream to and downstream of the cylinder. Again, it is illustrated how the velocity profiles are not influenced by the horizontal cylinder at the beginning of the area of interest, in all the three experimental runs. For Run 3 only, negative values of u are shown upstream to the cylinder, owing to the presence of the vortex originated from the interaction between the current and the obstacle. Immediately downstream of the cylinder in Run 1 and Run 2, the streamwise flow velocity is negative, indicating a reverse flow. Increasing the distance from the cylinder, it loses intensity and the vertical profiles start recovering the undisturbed upstream condition. However, the velocity profiles are different if we refer to Run 3, in which the shear layer separates from the cylinder, creating a recirculation zone.

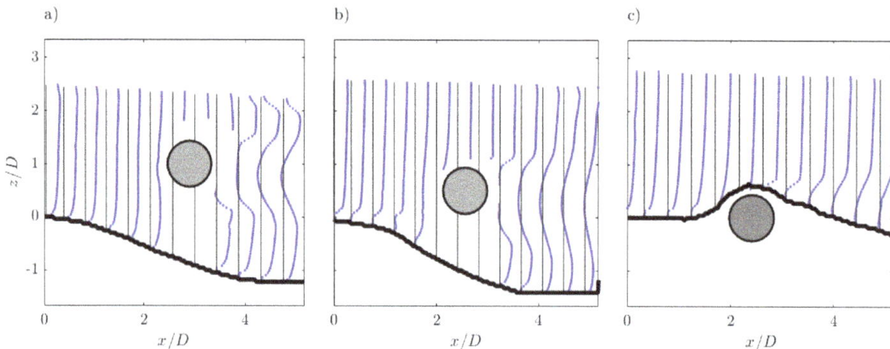

Figure 6. Vertical profiles of the ensemble-averaged streamwise flow velocity at different streamwise distances in (**a**) Run 1, (**b**) Run 2 and (**c**) Run 3.

3.3. Instantaneous and Ensemble-Averaged Vorticity Fields

The contours of dimensionless instantaneous vorticity $\omega_y D/U$ are shown in Figure 7 at different times t. Here, ω_y is the instantaneous vorticity given by $\partial u/\partial z - \partial w/\partial x$, whose positive and negative values specify clockwise and counterclockwise fluid motions, respectively. Specifically, the counterclockwise fluid motion causes the flow to accelerate, resulting in a downward transport of momentum in the downstream direction, and the clockwise fluid motion causes the flow to decelerate, resulting in an upward transport of momentum in the upstream direction [47].

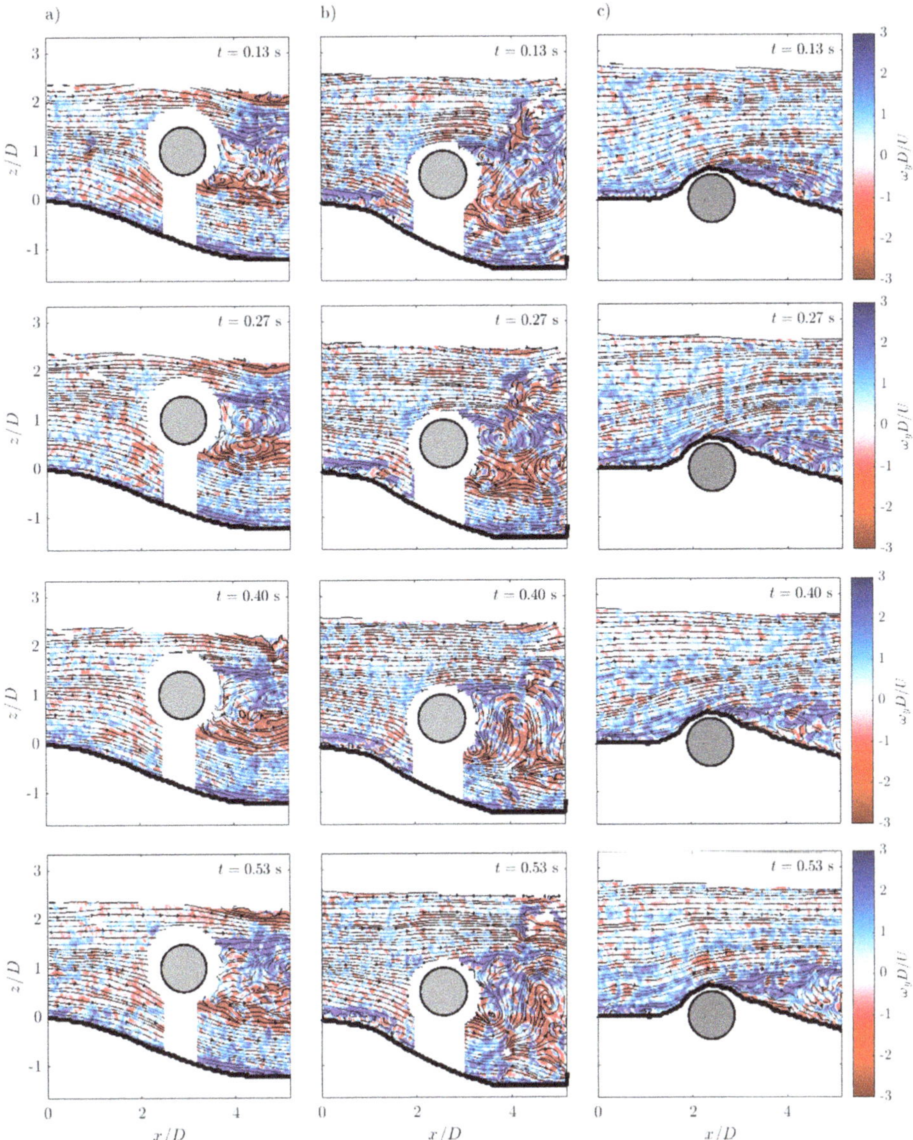

Figure 7. Dimensionless instantaneous vorticity fields at different times t on a vertical plane in (**a**) Run 1, (**b**) Run 2 and (**c**) Run 3.

From the analysis of Figure 7, the patterns of vorticity show that in the wake region of Run 1 and Run 2 a cluster of vorticity takes place along the shear layers. This is due to the separation induced by a pressure increase along the cylinder surface. The vortex that originates from the separation zone at the free-stream side of the cylinder has a positive sign indicating a clockwise rotation, whereas the vortex from the wall-side has a negative sign with a counterclockwise rotation. These motions are responsible for the production of the Reynolds shear stresses [48]. It is also evident that for Run 2 the steep slope of the scour hole behind the cylinder forces the shear layer originating from the lower edge of the cylinder to bend upwards, thus causing the associated lower vortex to interact with the upper one prematurely, as it was observed by Jensen et al. [29]. As regards to Run 3, the vortex downstream of the cylinder rotates with clockwise direction in the scour hole, creating a recirculation zone with decelerated fluid motion. However, in all the three cases it is found that the vortices do not change their rotational sense with time; therefore, they are recursive with time.

Figure 8 shows the streamlines and the contours of the dimensionless ensemble-averaged vorticity $\overline{\omega}_y D / U$, to substantiate the effects illustrated by the instantaneous vorticity patterns. Here, $\overline{\omega}_y$ is the ensemble-averaged vorticity given by $\partial \overline{u} / \partial z - \partial \overline{w} / \partial x$. At the upstream end of the investigated area, the vorticity is null along the water depth in all the three experimental runs. However, in the near-bed flow clockwise vortices occur owing to the bed roughness. For Run 1 and Run 2, downstream of the cylinder, in the wake region, vorticity assumes positive values in the upper zone and negative values in the lower part, that is with clockwise and counterclockwise rotations, respectively. Instead, Run 3 shows the clockwise vortex in the scour hole, as previously discussed.

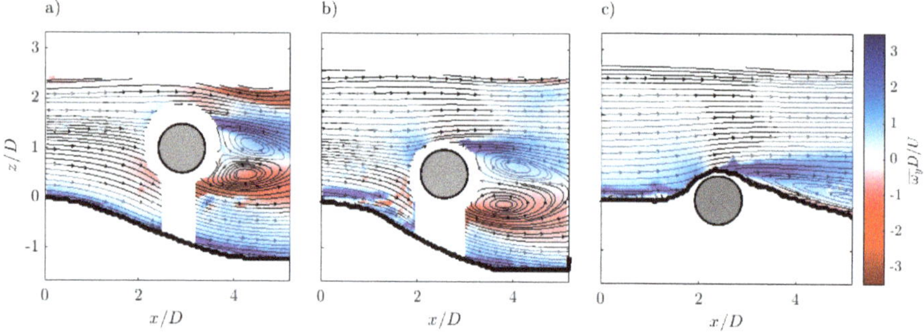

Figure 8. Dimensionless ensemble-averaged vorticity fields on a vertical plane in (**a**) Run 1, (**b**) Run 2 and (**c**) Run 3.

3.4. Viscous and Reynolds Stresses

The variations of the dimensionless viscous stresses τ_v in the flow domain, calculated as $(\nu d\overline{u}/dz)/u_*^2$ are depicted in Figure 9. Instead, the dimensionless Reynolds shear stresses $\tau_{uw} = -\overline{u'w'}/u_*^2$ are shown in Figure 10.

By comparing Figures 9 and 10, it is evident that upstream to the cylinder the viscous stresses are negligible along the water depth and increase in the near-bed flow zone in all the experimental runs. At the same time, τ_{uw} tends to assume a peak value at a certain distance from the bed surface and then starts decreasing. A different behavior is shown in Run 3, since a small region of negative values of τ_{uw} is evident close to the bed surface, owing to the small vortex on the upstream front of the cylinder. Considering Run 1 and Run 2, downstream of the cylinder two regions characterized by the maximum positive and negative values of the Reynolds shear stress occur. This suggests strong turbulent mixing process in these areas, indicating a downward transport of momentum in the downstream direction and an upward transport of momentum in the upstream direction, respectively. Close to the cylinder surface, τ_v assumes the highest positive and negative values where the shear stress is almost null.

As regards Run 3, both τ_v and τ_{uw} assume their maximum values in the scour hole, which suggests a strong turbulent mixing process that entrains high-momentum fluid into the recirculation region.

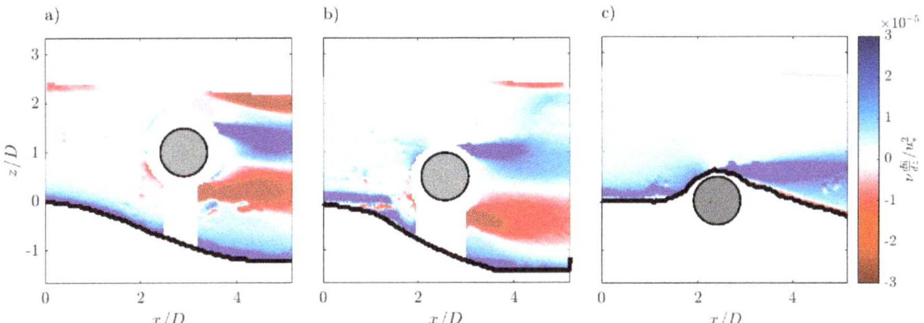

Figure 9. Dimensionless viscous stresses on a vertical plane in (**a**) Run 1, (**b**) Run 2 and (**c**) Run 3.

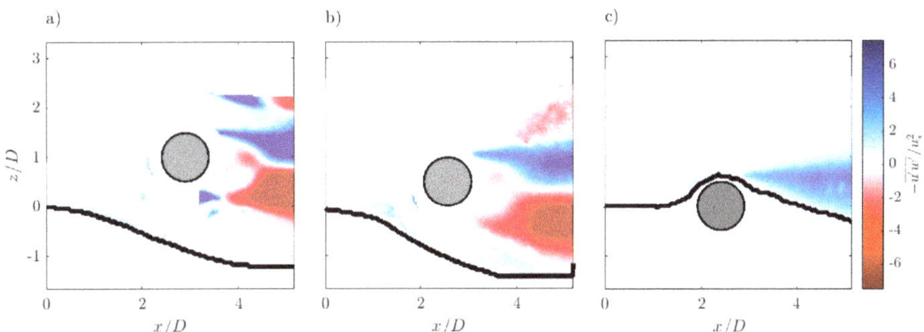

Figure 10. Dimensionless Reynolds shear stresses on a vertical plane in (**a**) Run 1, (**b**) Run 2 and (**c**) Run 3.

Finally, the dimensionless streamwise and vertical Reynolds normal stresses, expressed as $\sigma_{uu} = -\overline{u'u'}/u_*^2$ and $\sigma_{ww} = -\overline{w'w'}/u_*^2$, respectively, are shown in Figures 11 and 12.

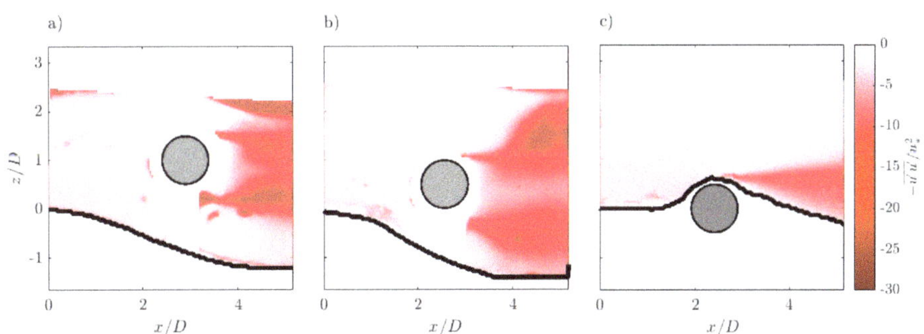

Figure 11. Dimensionless streamwise Reynolds normal stresses on a vertical plane in (**a**) Run 1, (**b**) Run 2, (**c**) Run 3.

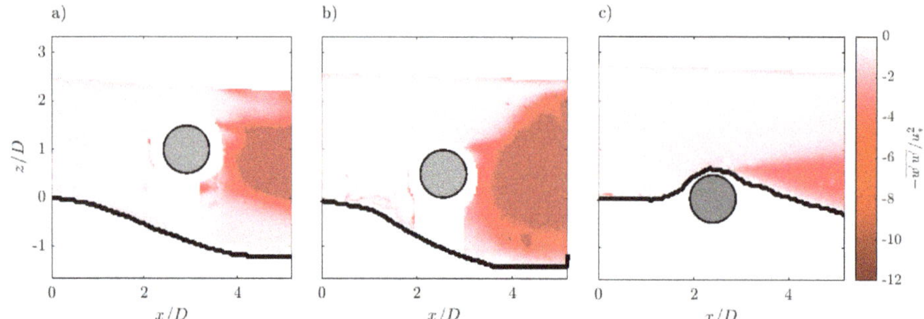

Figure 12. Dimensionless vertical Reynolds normal stresses on a vertical plane in (**a**) Run 1, (**b**) Run 2, (**c**) Run 3.

In all the experimental runs, upstream to the cylinder, it can be seen that σ_{uu} and σ_{ww} assume their maximum negative values in the near-bed flow region, and then they increase with z. This may be attributed to the fluid mixing that occurs in the presence of bed roughness. At the same time, it can be noted that downstream of the cylinder the intense fluid mixing induced by the presence of the obstacle enhances both the fluctuations u' and w'. The comparison between Run 1 and Run 2 shows that a smaller gap between the cylinder and the bed surface induces the flow area to enlarge with increased magnitude of both u' and w'.

On the other hand, Run 3 shows downstream of the cylinder a decrease of σ_{uu} and σ_{ww} in the recirculation zone, indicating, also in this case, fluid mixing with a maximum magnitude of 0.08 m/s and 0.05 m/s for u' and w', respectively, that are less than those in Run 1 and Run 2, for which u' and w' are about 0.14 m/s and 0.09 m/s. This leads to a decrease of about 65% for both σ_{uu} and σ_{ww}.

3.5. Ensemble-Averaged Turbulence Indicators

In order to measure the local turbulence level in a flow, the use of the ensemble-averaged turbulence indicator I may represent a useful tool [47]. This parameter can be estimated as follows:

$$I = \left(\frac{\frac{2}{3}k}{U^2} \right)^{0.5},$$

(2)

where k is the turbulent kinetic energy (TKE), expressed as:

$$k = \frac{1}{2}\left(\overline{u'u'} + \overline{v'v'} + \overline{w'w'} \right),$$

(3)

v' being the fluctuations of the spanwise velocity component with respect to its ensemble-averaged value. Since the PIV system allows the measurement of only two velocity components, the term $\overline{v'v'}$ can be approximated as $0.5\left(\overline{u'u'} + \overline{w'w'} \right)$ [49,50]. Therefore, the TKE is expressed as:

$$k = 0.75\left(\overline{u'u'} + \overline{w'w'} \right).$$

(4)

The turbulence level is high if $I > 0.5$, moderate if $0.1 < I < 0.5$ and low if $I < 0.1$ [51]. Thus, the contours of the ensemble-averaged turbulence indicator for the three experimental runs are shown in Figure 13. It is clear that for Runs 1 and 2 the magnitudes of I are moderate immediately downstream of the cylinder ($I \approx 0.3$), where the mean velocity is very low and a reverse flow takes place. In the wake region, the fluctuations in the streamwise and vertical directions cause the enhancement of the turbulence level, but it can be still classified as moderate ($I \approx 0.5$ for Run 1 and $I \approx 0.4$ for Run 2). However, the area characterized by the enhancement of the turbulence level is greater in Run 2 than in Run 1, since in this case it interests the entire flow depth. This can be ascribed to the fact that the increase of the gap between the cylinder and the bed surface (Run 1) causes the formation of two opposite vortices shorter and more compressed than those of Run 2. This also leads to higher I value in Run 1 than in Run 2. As regards Run 3, the turbulence level reaches a maximum value of about 0.4 in the scour hole downstream of the cylinder and, therefore, it can be classified as a moderate one. Nevertheless, in the rest of the flow domain, the turbulence level is basically low and it decreases as z increases, implying that the roughness effects and the influence of the cylinder diminishes owing to the damping in u' and w', which is visible also in Figures 11 and 12.

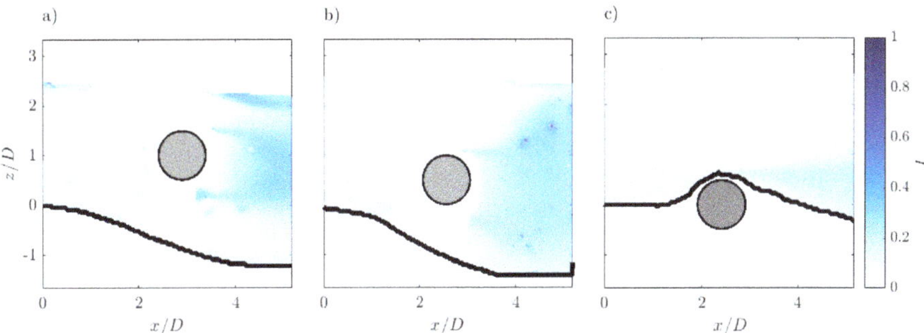

Figure 13. Ensemble-averaged turbulence level on a vertical plane in (**a**) Run 1, (**b**) Run 2 and (**c**) Run 3.

4. Conclusions

This study examines the turbulence characteristics at scoured horizontal cylinders in different condition of submergence, in order to deepen the knowledge investigating the flow-structure interactions in the presence of a mobile bed in shallow-water condition. Three experimental runs were performed considering a suspended cylinder, a laid on cylinder and a half-buried cylinder. Specifically, bed scouring occurred in all the runs and the resulting bed configurations were analyzed in order to better comprehend the flow dynamics.

For the first two conditions, at the end of the scouring phase, it was demonstrated that the incoming flow approaches the upstream face of the cylinder and causes flow separation. Thus, the main flow is divided into two flows: one part is oriented towards the bed surface, whereas the other part is oriented towards the free surface. The velocity at the underside of the cylinder is slightly lower than the one at the top side, as observed by Jensen et al. [29] in their experiments. This latter is also responsible for the formation of the scour hole that involves the area upstream to and beneath the cylinder owing to tunnel erosion. This phenomenon is more pronounced when the cylinder is laid on the bed. In the wake zone, for the suspended cylinder a greater amount of flow passes below it. Therefore, this flow tends to expand immediately downstream of the cylinder, limiting the extension of the induced vortex system. The analysis of the vorticity field revealed that in the wake region vorticity assumes positive values in the upper zone and negative values in the lower part, that is with clockwise and counterclockwise rotations, respectively. This is in accordance with the results obtained by Bearman and Zdravkovich [21], Price et al. [24] and others, since it was demonstrated that vortex shedding occurs at gaps higher than 0.3D. The strong turbulent mixing process in these areas is highlighted by the Reynolds shear stress distribution, indicating a downward transport of momentum in the downstream direction and an upward transport of momentum in the upstream direction, respectively. To support these observations, the turbulence level was calculated and it was found that fluctuations in the streamwise and vertical directions cause the enhancement of the turbulence level immediately downstream of the obstacle, reaching a value within the range 0.3–0.5, indicating a moderate turbulence level.

A different behavior was observed when the cylinder was initially half-buried. Tunnel erosion did not occur and, as a consequence, the obstruction represented by the cylinder created an adverse pressure gradient resulting in the separation of flow lines forming a vortex on the upstream front of the cylinder. Separation of flow also took place at the free-stream side of the cylinder. A reverse roller was formed in the wake region. This was responsible for the local scour downstream of the cylinder. The vorticity pattern showed the presence of the clockwise vortex in the scour hole with a strong mixing process that entrains high-momentum fluid into the recirculation region. It was demonstrated that here the turbulence level can be classified as moderate ($I \approx 0.4$).

Additional work in the future is required to take into account other variables that may affect the investigated turbulence structures (e.g., cylinder diameter, shape of the pipe, flow discharge, water depth), their development during the scouring process and also other in-depth statistical analyses (such as the third-order statistics to provide a very accurate measure of the TKE dissipation rate [52] or correlation functions to analyze the turbulent coherent structures [53]).

This study represents an advancement in the current understanding of the flow-structure interaction at scoured horizontal cylinders subjected to currents. It was demonstrated that bed changes should be considered in the analysis of the flow field and in the prediction of the forces acting on the cylinder rather than considering a plane boundary, because the effects are very different and depend also on the gap between the cylinder and the bed itself.

Author Contributions: Conceptualization, N.P. and R.G.; Data curation, N.P. and F.C.; Formal analysis, N.P. and F.C.; Methodology, N.P., F.C. and R.G.; Writing—Original Draft Preparation, N.P.; Writing—Review and Editing, N.P., F.C. and R.G.; Supervision, R.G. All authors have read and agreed to the published version of the manuscript.

Funding: This research received no external funding.

Acknowledgments: The authors would like to thank Antonio Leuzzi for his valuable work during the performance of the experimental Runs.

Conflicts of Interest: The authors declare no conflict of interest.

References

1. Fredsøe, J. Pipeline-seabed interaction. *J. Waterw. Port Coast. Ocean Eng.* **2016**, *142*, 03116002. [CrossRef]
2. Azamathulla, H.M.; Yusoff, M.A.M.; Hasan, Z.A. Scour below submerged skewed pipeline. *J. Hydrol.* **2014**, *509*, 615–620. [CrossRef]
3. Chiew, Y.M. Prediction of maximum scour depth at submarine pipelines. *J. Hydraul. Eng.* **1991**, *117*, 452–466. [CrossRef]
4. Dey, S.; Singh, N.P. Clear-water scour below underwater pipelines under steady flow. *J. Hydraul. Eng.* **2008**, *134*, 588–600. [CrossRef]
5. Chao, J.L.; Hennessy, P.V. Local scour under ocean outfall pipelines. *J. Water Pollut. Control Fed.* **1972**, *44*, 1443–1447.
6. Herbich, J.B. Scour around pipelines and other objects. In *Offshore Pipeline Design Elements*; Marcell Dekker, Inc.: New York, NY, USA, 1981; pp. 43–96.
7. Herbich, J.B. Hydromechanics of submarine pipelines: Design problems. *Can. J. Civ. Eng.* **1985**, *12*, 863–874. [CrossRef]
8. Herbich, J.B.; Schiller, R.E., Jr.; Watanabe, R.K.; Dunlap, W.A. Scour around pipelines. In *Sea Floor Scour—Design Guidelines for Ocean Founded Structures*; Marcell Dekker, Inc.: New York, NY, USA, 1984; pp. 203–210.
9. Kjeldsen, S.P.; Gjorsvik, O.; Bringaker, K.G.; Jacobsen, J. Local scour near offshore pipelines. In Proceedings of the Second International Conference on Port and Ocean Engineering Under Arctic Conditions (POAC), Reykjavík, Iceland, 27–30 August 1973.
10. Bijker, E.W.; Leeuwestein, W. Interaction between pipelines and the seabed under the influence of waves and currents. In *Seabed Mechanics*; Springer: Dordrecht, The Netherlands, 1984; pp. 235–242.
11. Ibrahim, A.; Nalluri, C. Scour prediction around marine pipelines. In Proceedings of the 5th International Symposium on Offshore Mechanics and Arctic Engineering, Tokyo, Japan, 13 April 1986; pp. 679–684.

12. Hansen, E.A.; Fredsøe, J.; Mao, Y. Two-dimensional scour below pipelines. In Proceedings of the 5th International Symposium on Offshore Mechanics and Arctic Engineering, Tokyo, Japan, 13–18 April 1985; pp. 670–678.

13. Mao, Y. *The Interaction between a Pipeline and an Erodible Bed*; Institute of Hydrodynamics and Hydraulic Engineering, Technical University of Denmark: Lyngby, Denmark, 1986; p. 39.

14. Maza, J.A. Introduction to river engineering. In *Advanced Course on Water Resources Management*; Universitá Italiana per Stranieri: Perugia, Italy, 1987.

15. Moncada-M., A.T.; Aguirre-Pe, J. Scour below pipeline in river crossings. *J. Hydraul. Eng.* **1999**, *125*, 953–958. [CrossRef]

16. Chiew, Y.M. Mechanics of local scour around submarine pipelines. *J. Hydraul. Eng.* **1990**, *116*, 515–529. [CrossRef]

17. Li, F.; Cheng, L. Numerical model for local scour under offshore pipelines. *J. Hydraul. Eng.* **1999**, *125*, 400–406. [CrossRef]

18. Brørs, B. Numerical modeling of flow and scour at pipelines. *J. Hydraul. Eng.* **1999**, *125*, 511–523. [CrossRef]

19. Mohr, H.; Draper, S.; Cheng, L.; White, D.J. Predicting the rate of scour beneath subsea pipelines in marine sediments under steady flow conditions. *Coast. Eng.* **2016**, *110*, 111–126. [CrossRef]

20. Gao, F.P.; Yang, B.; Wu, Y.X.; Yan, S.M. Steady current induced seabed scour around a vibrating pipeline. *Appl. Ocean Res.* **2006**, *28*, 291–298. [CrossRef]

21. Bearman, P.W.; Zdravkovich, M.M. Flow around a circular cylinder near a plane boundary. *J. Fluid Mech.* **1978**, *89*, 33–47. [CrossRef]

22. Zdravkovich, M.M. Intermittent flow separation from flat plate induced by a nearby circular cylinder. In Proceedings of the International Symposium on Flow Visualization, Bochum, Germany, 9–12 September 1981; pp. 219–224.

23. Lei, C.; Cheng, L.; Kavanagh, K. Re-examination of the effect of a plane boundary on force and vortex shedding of a circular cylinder. *J. Wind Eng. Ind. Aerodyn.* **1999**, *80*, 263–286. [CrossRef]

24. Price, S.J.; Sumner, D.; Smith, J.G.; Leong, K.; Paidoussis, M.P. Flow visualization around a circular cylinder near to a plane wall. *J. Fluids Struct.* **2002**, *6*, 175–191. [CrossRef]

25. Hatipoglu, F.; Avci, I. Flow around a partly buried cylinder in a steady current. *Ocean Eng.* **2003**, *30*, 239–249. [CrossRef]

26. Akoz, M.S.; Sahin, B.; Akilli, H. Flow characteristic of the horizontal cylinder placed on the plane boundary. *Flow Meas. Instrum.* **2010**, *21*, 476–487. [CrossRef]

27. Arslan, T.; Malavasi, S.; Pettersen, B.; Andersson, H.I. Turbulent flow around a semi-submerged rectangular cylinder. *J. Offshore Mech. Arct. Eng.* **2013**, *135*, 041801. [CrossRef]

28. Malavasi, S.; Guadagnini, A. Interactions between a rectangular cylinder and a free-surface flow. *J. Fluids Struct.* **2007**, *23*, 1137–1148. [CrossRef]

29. Jensen, B.L.; Sumer, B.M.; Jensen, H.R.; Fredsøe, J. Flow around and forces on a pipeline near a scoured bed in steady current. *J. Offshore Mech. Arct. Eng.* **1990**, *112*, 206–213. [CrossRef]

30. Smith, H.D.; Foster, D.L. Modeling of flow around a cylinder over a scoured bed. *J. Waterw. Port Coast. Ocean Eng.* **2005**, *131*, 14–24. [CrossRef]

31. Nezu, I.; Nakagawa, H. *Turbulence in Open-Channel Flows*, Balkema Publishers/IAHR–monograph: Rotterdam, The Netherlands, 1993; p. 293.

32. Neill, C.R. Mean-velocity criterion for scour of coarse uniform bed material. In Proceedings of the International Association of Hydraulic Research 12th Congress, Fort Collins, CO, USA, 11–14 September 1967; Volume 3, pp. 46–54.

33. Da Silva, A.F.; Yalin, M.S. *Fluvial Processes*, 2nd ed.; IAHR–monograph: Delft, The Netherlands, 2017; p. 266.

34. Padhi, E.; Penna, N.; Dey, S.; Gaudio, R. Hydrodynamics of water-worked and screeded gravel beds: A comparative study. *Phys. Fluids* **2018**, *30*, 085105. [CrossRef]

35. Balakumar, B.J.; Orlicz, G.C.; Tomkins, C.D.; Prestridge, K.P. Simultaneous particle-image velocimetry–planar laser-induced fluorescence measurements of Richtmyer–Meshkov instability growth in a gas curtain with and without reshock. *Phys. Fluids* **2008**, *20*, 124103. [CrossRef]

36. Balakumar, B.J.; Prestridge, K.P.; Orlicz, G.; Balasubramanian, S.; Tomkins, C. High resolution experimental measurements of Richtmyer-Meshkov turbulence in fluid layers after reshock using simultaneous PIV-PLIF. *AIP Conf. Proc.* **2009**, *1195*, 659–662. [CrossRef]

37. Gonçalves, J.A.; Henriques, R. UAV photogrammetry for topographic monitoring of coastal areas. *ISPRS J. Photogramm. Remote Sens.* **2015**, *104*, 101–111. [CrossRef]

38. Fortunato, G.; Funari, M.F.; Lonetti, P. Survey and seismic vulnerability assessment of the Baptistery of San Giovanni in Tumba (Italy). *J. Cult. Herit.* **2017**, *26*, 64–78. [CrossRef]

39. Bertin, S.; Friedrich, H.; Delmas, P.; Chan, E.; Gimel'farb, G. Dem quality assessment with a 3d printed gravel bed applied to stereo photogrammetry. *Photogramm. Rec.* **2014**, *29*, 241–264. [CrossRef]

40. Penna, N.; D'Alessandro, F.; Gaudio, R.; Tomasicchio, G.R. Three-dimensional analysis of local scouring induced by a rotating ship propeller. *Ocean Eng.* **2019**, *188*. [CrossRef]

41. Sumer, B.M.; Fredsøe, J. *The Mechanics of Scour in the Marine Environment*; Advanced Series on Ocean Engineering; World Scientific: Singapore, 2002; Volume 17, p. 552.

42. Sumer, B.M.; Truelsen, C.; Sichmann, T.; Fredsøe, J. Onset of scour below pipelines and self-burial. *Coast. Eng.* **2001**, *42*, 313–335. [CrossRef]

43. Sumer, B.M.; Fredsøe, J. *Hydrodynamics around Cylindrical Structures*; Advanced Series on Ocean Engineering; World Scientific: Singapore, 2006; Volume 26, p. 548.

44. Grass, A.J.; Raven, P.W.J.; Stuart, R.J.; Bray, J.A. The influence of boundary layer velocity gradients and bed proximity on vortex shedding from free spanning pipelines. *J. Energy Resour. Technol.* **1984**, *106*, 70–78. [CrossRef]

45. Raven, P.W.J.; Stuart, R.J.; Bray, J.A.; Littlejohns, P.S. Full-scale dynamic testing of submarine pipeline spans. In Proceedings of the 17th Annual Offshore Technology Conference, Houston, TX, USA, 6–9 May 1985; paper No. 5005. pp. 395–404.

46. Roshko, A. Experiments on the flow past a circular cylinder at very high Reynolds number. *J. Fluid Mech.* **1961**, *10*, 345–356. [CrossRef]

47. Padhi, E.; Penna, N.; Dey, S.; Gaudio, R. Near-bed turbulence structures in water-worked and screeded gravel-bed flows. *Phys. Fluids* **2019**, *31*, 045107. [CrossRef]

48. Adrian, R.J. Structure of turbulent boundary layers. In *Coherent Flow Structures at Earth's Surface*; Venditti, J.G., Best, J.L., Church, M., Hardy, R.J., Eds.; John Wiley & Sons: Chichester, UK, 2013; pp. 17–24.

49. Antonia, R.A.; Luxton, R.E. The response of a turbulent boundary layer to a step change in surface roughness. Part 2. Rough-to-smooth. *J. Fluid Mech.* **1972**, *53*, 737–757. [CrossRef]

50. Padhi, E.; Penna, N.; Dey, S.; Gaudio, R. Spatially averaged dissipation rate in flows over water-worked and screeded gravel beds. *Phys. Fluids* **2018**, *30*, 125106. [CrossRef]

51. Russo, F.; Basse, N.T. Scaling of turbulence intensity for low-speed flow in smooth pipes. *Flow Meas. D Instrum.* **2016**, *52*, 101–114. [CrossRef]

52. Coscarella, F.; Servidio, S.; Ferraro, D.; Carbone, V.; Gaudio, R. Turbulent energy dissipation rate in a tilting flume with a highly rough bed. *Phys. Fluids* **2017**, *29*, 085101. [CrossRef]

53. Caroppi, G.; Västilä, K.; Järvelä, J.; Rowiński, P.M.; Giugni, M. Turbulence at water-vegetation interface in open channel flow: Experiments with natural-like plants. *Adv. Water Resour.* **2019**, *127*, 180–191. [CrossRef]

Article

Anisotropy in the Free Stream Region of Turbulent Flows through Emergent Rigid Vegetation on Rough Beds

Nadia Penna *, Francesco Coscarella, Antonino D'Ippolito and Roberto Gaudio

Dipartimento di Ingegneria Civile, Università della Calabria, 87036 Rende (CS), Italy;
francesco.coscarella@unical.it (F.C.); antonino.dippolito@unical.it (A.D.); gaudio@unical.it (R.G.)
* Correspondence: nadia.penna@unical.it; Tel.: +39-0984-496-553

Received: 10 August 2020; Accepted: 30 August 2020; Published: 2 September 2020

Abstract: Most of the existing works on vegetated flows are based on experimental tests in smooth channel beds with staggered-arranged rigid/flexible vegetation stems. Actually, a riverbed is characterized by other roughness elements, i.e., sediments, which have important implications on the development of the turbulence structures, especially in the near-bed flow zone. Thus, the aim of this experimental study was to explore for the first time the turbulence anisotropy of flows through emergent rigid vegetation on rough beds, using the so-called anisotropy invariant maps (AIMs). Toward this end, an experimental investigation, based on Acoustic Doppler Velocimeter (ADV) measures, was performed in a laboratory flume and consisted of three runs with different bed sediment size. In order to comprehend the mean flow conditions, the present study firstly analyzed and discussed the time-averaged velocity, the Reynolds shear stresses, the viscous stresses, and the vorticity fields in the free stream region. The analysis of the AIMs showed that the combined effect of vegetation and bed roughness causes the evolution of the turbulence from the quasi-three-dimensional isotropy to axisymmetric anisotropy approaching the bed surface. This confirms that, as the effects of the bed roughness diminish, the turbulence tends to an isotropic state. This behavior is more evident for the run with the lowest bed sediment diameter. Furthermore, it was revealed that also the topographical configuration of the bed surface has a strong impact on the turbulent characteristics of the flow.

Keywords: anisotropy; rigid vegetation; sediments; turbulent flow

---

1. Introduction

The flow through emergent rigid vegetation has been widely investigated by researchers, both experimentally and numerically, aiming at analyzing the effects of vegetation on the flow structure and its implications on hydraulic resistance, turbulent structures, mixing processes, and sediment transport [1–5]. This particular type of vegetation (rigid cylinders) can simulate rigid reeds or trees in riparian environments [6,7], when the flow does not hit the foliage, since the dynamic plant motions exhibited by real vegetation is neglected [8].

Most of the existing works were primarily conducted on smooth channel beds with staggered-arranged vegetation stems (e.g., [1,2,6,9–16]). However, special interest must be devoted to studies on vegetated flows with rough beds, since the interactions between flow, vegetation, and sediments permit achieving a better understanding of the turbulence structures in natural environments, which have a key role in the sediment transport mechanism. In fact, the flow and turbulence characteristics through emergent rigid vegetation on rough beds are still poorly understood. As reported by Maji et al. [17], flow conditions and the solid volume fraction of emergent vegetation affect the individual contributions of sweep and ejection coherent structures, which play a dominant role in dislodging bed particles. Therefore, an in-depth description of turbulent structure is imperative for the correct understanding of the sediment transport process and for the development of new sediment transport theories in emergent vegetated flows.

Recently, Penna et al. [18] analyzed the flow field, the turbulent kinetic energy (TKE), and the energy spectra of velocity fluctuations around a rigid stem on three different rough beds. They showed that, in the region below the free surface region, the flow is strongly influenced by the vegetation. Moving toward the bed, the flow is affected by a combined effect of both vegetation and bed roughness. At the same time, Penna et al. [18] noted a strong lateral variation of TKE from the flume centerline to the cylinder in the intermediate region. Finally, the analysis of the energy spectra revealed that, in the near-bed flow region at low wave numbers, the macro-turbulence is governed by the bed roughness, regardless of the measurement point location with respect to the vegetation stem. In the region of wake vortexes (i.e., downstream of the vegetation stem), the macro-turbulence is extended at smaller scales, implying a strong influence of the vegetation.

Nevertheless, a comprehensive characterization of the turbulence structures in the different flow layers cannot disregard from the turbulence anisotropy investigation. In fact, one of the most frequently analyzed quantities in turbulence studies is the anisotropic behavior of turbulence, in terms of the degree of departure from the isotropic turbulence. This is a common feature of complex fluid flows [19], such as those that characterize vegetated channels. The 'isotropic turbulence' refers to an idealized condition, in which the velocity fluctuations do not vary regardless of the rotation of axes [20]. This means that the Reynolds normal stresses along the streamwise, spanwise, and vertical directions (σ_{uu}, σ_{vv}, and σ_{ww}, respectively) are the same. Conversely, in the 'anisotropic turbulence' the Reynolds normal stresses cannot be considered as invariant, because the temporal velocity fluctuations along the three axes follow a preferential direction [20].

To characterize the type of turbulence, one of the most used methodologies is the definition of the Reynolds stress anisotropy tensor. In particular, the diagonalization of the tensor provides three eigenvalues (λ_1, λ_2, and λ_3) and three eigenvectors (e_1, e_2, and e_3) of the turbulence anisotropy [19]. The anisotropy invariant map (AIM) describes the domain of all potential turbulent flows considering the second and third invariants. In fact, it is a 2D domain with a triangular shape, whose boundaries are characteristic of turbulence state (1D, 2D, and 3D turbulence) and related processes (axisymmetric expansion, axisymmetric contraction, and two-component turbulence) [3].

Hence, the driving idea of the present study was the description of the turbulence anisotropy (with the AIMs) of flows through emergent rigid vegetation on rough beds. Indeed, exploring for the first time this crucial aspect in vegetated flows may advance the current understanding of the flow–vegetation–roughness interaction by describing the evolution of the stress ellipsoid formed by the Reynolds stresses. To this end, an experimental campaign was performed in a uniformly vegetated channel varying the bed sediment size (coarse sand, fine gravel, and coarse gravel), under the same hydraulic conditions. Additionally, in order to better comprehend the overall flow conditions, the time-averaged velocity, the Reynolds shear stresses, the viscous stresses, and the vorticity fields were analyzed and briefly discussed.

2. Experimental Program

2.1. Flume Set-Up and Bed Sediments

The experimental study was performed in a 9.6 m long, 0.485 m wide, and 0.5 m deep flow recirculating tilting flume at the *Laboratorio "Grandi Modelli Idraulici"* (GMI), *Università della Calabria*, Rende, Italy. Three experimental runs were performed under the same approach flow conditions and with the same vegetation arrangement, but with different bed sediments. In particular, the flume bed was covered with a 20 cm thick layer of uniform, very coarse sand ($d_{50} = 1.53$ mm), fine gravel ($d_{50} = 6.49$ mm), and coarse gravel ($d_{50} = 17.98$ mm) for Runs 1, 2, and 3, respectively. Figure 1 shows the grain-size distribution of the mixtures used to create the bed, which were obtained by analyzing three samples for each Run.

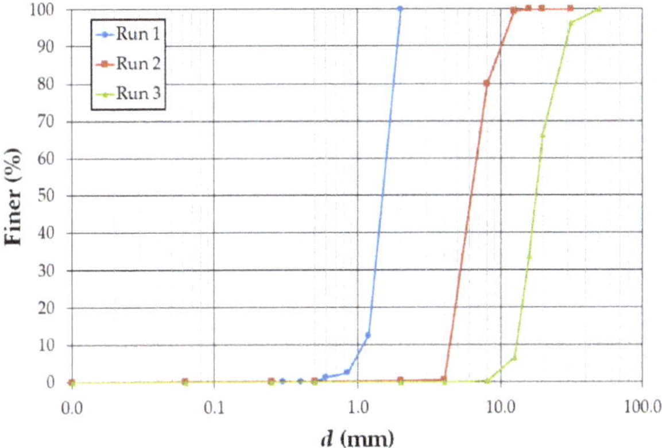

Figure 1. Grain-size distribution curve of the bed sediments for each experimental run.

The flow rate Q was controlled by a submerged pump. It was measured using a calibrated sharp-crested V notch weir installed in a downstream tank, where the outflow was collected. To reduce the disturbance at the flume entrance and to dampen the related turbulence level, honeycombs with a diameter of 10 mm were used. The flume slope was set equal to 1.5‰ by maneuvering a hydraulic jack. Furthermore, the flow depth h was regulated with a downstream tailgate and measured with a point gauge.

The experimental runs were carried out by using a uniformly distributed vegetated channel bed, where the emergent vegetation was simulated with vertical, rigid, and circular wooden cylinders. A total number of 68 cylinders, each 2 cm in diameter, were inserted into a 1.96 m long, 0.485 m wide, and 0.015 m thick Plexiglas panel, which in turn was fixed to the channel bottom. The test section was located 6 m downstream of the flume inlet. The cylinders were arranged in an aligned pattern where the axis-to-axis distance between the stems was equal to 12 cm in both the streamwise and spanwise directions. Therefore, the total number of stems per unit area was 71.6 m^{-2}. The frontal area per canopy volume was $a = d/\Delta S^2 = 1.4$ m^{-1}, where d is the stem diameter and ΔS the axis-to-axis distance between the stems. The solid volume fraction occupied by the stems was $\phi = (\pi/4)ad = 0.02$; thus, the vegetation can be considered as dense [21]. Further details of the experimental setup were recently reported by Penna et al. [18].

2.2. Experimental Procedure

Initially, the flume was filled in with sediments and was subsequently screeded flat to obtain a bed with a mean surface elevation having the same longitudinal slope of the channel. All the experimental runs initiated with a steady flow rate equal to 19.73 l s^{-1} and a water depth of 14 cm (measured 50 cm upstream to the vegetation array), designed to prevent sediments motions and to satisfy the fixed bed condition. Thus, the average flow velocity U was 0.30 m s^{-1} (=$Q/(Bh)$, where B is the flume width) and the flow Froude number Fr was 0.26 (=$U/(gh)^{0.5}$, where g is the gravitational acceleration). For each run, Table 1 shows the details of the experimental conditions for the approaching flow (measured 50 cm upstream to the test section), where u_* is the shear velocity determined extending the linear trend of the Reynolds shear stress (RSS) distribution down to the maximum crest level (=$\left(-\overline{u'w'}\right)^{0.5}$, where u' and w' are the temporal velocity fluctuations in the streamwise and vertical directions, respectively, and the symbol ‾ indicates the time average), T is the water temperature (measured with an integrated thermometer with an accuracy of 0.1 °C) and v is the water kinematic viscosity, determined as a function of the water temperature [22]. Furthermore, the Reynolds number of the sediments Re_* (=$u_*\varepsilon/v$, where

ε is the equivalent Nikuradse sand roughness height, equal to about 2d_{50}) and the Reynolds number of the vegetation stems Re_d (=Ud/v, where d is the stem diameter) were calculated.

Table 1. Details of the experimental conditions.

Run	d_{50} (mm)	u_* (m s^{-1})	T (°C)	v (m^2 s)	Re_*	Re_d
1	1.53	0.017	21.70	9.63×10^{-7}	54	6231
2	6.49	0.022	21.44	9.69×10^{-7}	295	6192
3	17.98	0.028	20.80	9.83×10^{-7}	1024	6104

The instantaneous three-dimensional flow velocity components were measured with a down-looking Vectrino probe (Acoustic Doppler Velocimeter, ADV) manufactured by Nortek, Vangkroken, Norway. The measurements were performed along the flume centerline at various relative streamwise distances x/L_s = 0, 0.17, 0.33, 0.50, 0.67, 0.83, 1.00, where x is the streamwise direction and L_s is the length of the study area equal to 12 cm. Note that x/L_s = 0 is the origin of the study area, located 6.78 m downstream of the channel entrance, and x/L_s = 0.50 corresponds to the vegetation stem axis. The 3D velocity components (u, v, w) refer to (x, y, z), where y and z are the spanwise and vertical direction, respectively.

The ADV probe was installed on a motorized 3-axis traverse system (HR Wallingford Ltd., Wallingford, Oxfordshire, UK) to easily move the probe within the study area during the experimental run. The Vectrino was operated with a transmitting length of 0.3 mm and a sampling volume constituted of a cylinder of 6 mm in diameter and 1 mm high. The sampling duration was equal to 180 s (the sampling frequency was fixed to 100 Hz), assuring statistically time-independent time-averaged velocities and turbulence quantities. It was not possible to perform velocity measurements within the flow zone 5 cm below the free surface, because the ADV beams converge at 5 cm below the probe. Thus, the vertical resolutions were 3 mm for $z \leq$ 15 mm and 5 mm above, where z is the vertical axis starting from the maximum crest elevation in the study area.

The ADV data were pre-processed for detecting potential spikes with the phase-space thresholding method. Spikes were replaced with a third-order polynomial through 12 points on both sides of the spike itself, as suggested by Goring and Nikora [23].

3. Results and Discussion

3.1. Time-Averaged Flow

The dimensionless time-averaged velocity fields and 2D velocity vectors, having magnitude $\hat{u} = \left(\bar{u}^2 + \bar{w}^2\right)^{0.5}/u_*$ (where $\bar{u}$ and $\bar{w}$ are the time-averaged velocity components) and direction $\tan^{-1}(\bar{w}/\bar{u})$, on the vertical central plane are illustrated in Figure 2 for each Run. Here, the horizontal axis is represented as $\hat{x} = x/L_s$, and the vertical axis $\hat{z}$ was made dimensionless dividing z by the local flow depth h_l.

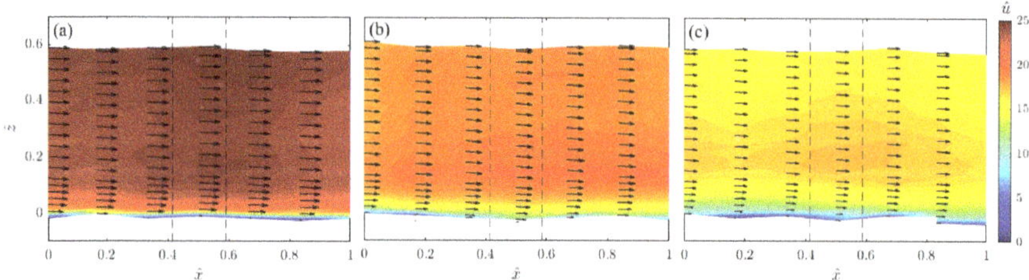

Figure 2. Contours of dimensionless time-averaged velocity and 2D velocity vectors measured on the vertical central plane for (**a**) Run 1, (**b**) Run 2, and (**c**) Run 3. The black broken lines indicate the edge of the vegetation stem.

In Figure 2, a streamwise variation of the velocity field is detected in the three Runs. Specifically, in correspondence of the vegetation stem, $\hat{u}$ increases with respect to the areas upstream to and downstream of the cylinder. This denotes the presence of: (1) a convergent flow between two stems and toward the flume centerline [24]; (2) a retarded flow owing to a divergent flow downstream of the stems. The changes of magnitude and direction of velocity vectors suggest the presence of a near-bed flow heterogeneity, which are more pronounced looking from Run 3 to Run 1. The streamwise and vertical variations of the velocity field is in agreement with Maji et al. [17], since significant velocity gradients were found in all the experimental runs. This is due to the different bed roughness that characterizes the three experimental runs. In essence, the flow velocity increases with the vertical distance, reaching the maximum values in correspondence of the vegetation stem and at the elevation $z \approx 0.15h_l$, regardless of the bed roughness, implying that this region is mainly influenced by the presence of vegetation. Here, the production of turbulence by the canopy exceeds the production by the bed shear [21].

Figure 3 presents the contours of the dimensionless Reynolds shear stresses $\hat{\tau}_{uw}$ $(= -\overline{u'w'}/u_*^2)$ on the vertical central plane for the three experimental Runs. They exhibit small magnitudes in the flow area dominated by the vegetation (for $z > 0.15h_l$) [6]. Moving toward the bed surface, the Reynolds shear stresses increase with a high gradient, owing to the bed roughness. As d_{50} increases, this zone becomes more extended. However, close to the bed at the crest level, the Reynolds shear stresses become negligible. The contours of $\hat{\tau}_{uw}$ reveal that they are not influenced by the position of vegetation stems, since their spatial distribution is quite uniform. Indeed, this agrees with the findings of Ricardo et al. [6], who demonstrated that the Reynolds stresses are not sensitive to local spatial gradients of the stem distribution, because they depend on the local number of stems per unit area. Analogous patterns can be noticed in Figure 4, where the dimensionless viscous stresses $\hat{\tau}_v$ $(= \nu(d\overline{u}/dz)/u_*^2)$ on the vertical central plane for the three experimental Runs are presented. In fact, the streamwise distribution of $\hat{\tau}_v$ is almost uniform in each Run. As the roughness decreases, the viscous shear stress increases at the crest level. Then, it diminishes as the vertical distance z increases. This agrees with the findings of Nepf [21], who stated that the viscous stress is negligible with respect to the vegetative drag over most of the depth, excluding a thin layer near the bed of a scale comparable to the stem diameter.

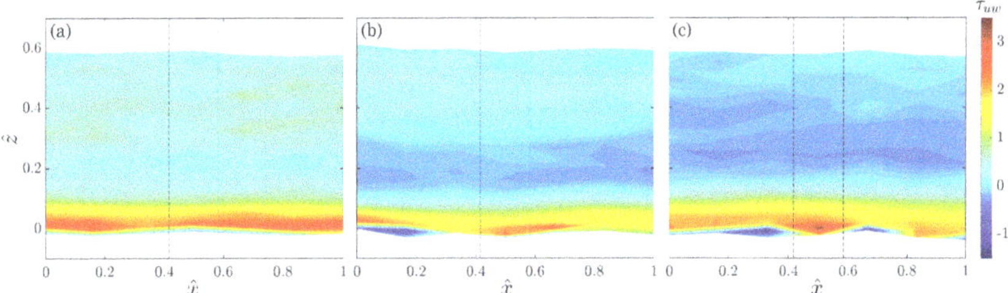

Figure 3. Contours of dimensionless Reynolds shear stresses measured on the vertical central plane for (a) Run 1, (b) Run 2, and (c) Run 3. The black broken lines indicate the edge of the vegetation stem.

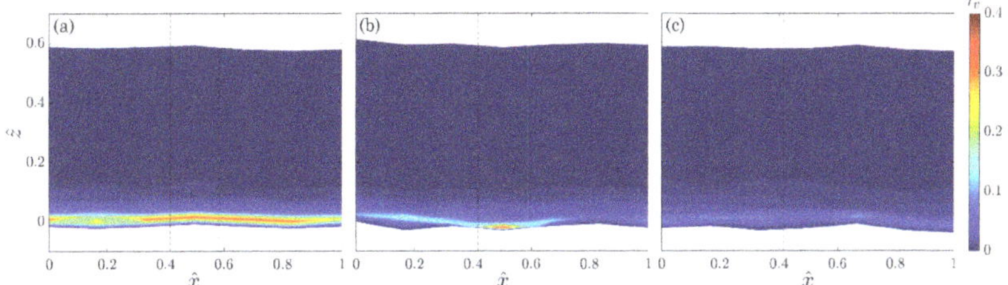

Figure 4. Contours of dimensionless viscous shear stresses measured on the vertical central plane for (a) Run 1, (b) Run 2, and (c) Run 3. The black broken lines indicate the edge of the vegetation stem.

The effects of the bed roughness structures were investigated through the analysis of the dimensionless vorticity of the time-averaged flow $\overline{\omega}_y d/u_*$ on the vertical central plane (Figure 5). Here, $\overline{\omega}_y$ is the vorticity of the time-averaged flow, given by $\partial\overline{u}/\partial z - \partial\overline{w}/\partial x$. Positive values of the vorticity indicate clockwise fluid motion; on the contrary, negative values refer to counterclockwise direction. Specifically, the rotational direction provides information about flow acceleration and deceleration in the near-bed flow: counterclockwise rotation induces flow acceleration, with a downward transport of momentum in the downstream direction; clockwise rotation causes flow deceleration, with upward transport of momentum in the upstream direction [25,26]. It is evident that the vorticity changes its signs alternatively in the flow layer affected by the presence of both vegetation and bed roughness. This implies the heterogeneity of the time-averaged near-bed flow: fluid streaks move alternatively in both clockwise and counterclockwise directions [25]. Furthermore, it is possible to note that the changes of the vorticity rotational direction are more frequent in Run 1 than in the other two runs along the streamwise direction. As observed by Ricardo et al. [27], the cylinders induce a regular structure of vortex patterns independently from the space between cylinders also in the horizontal plane.

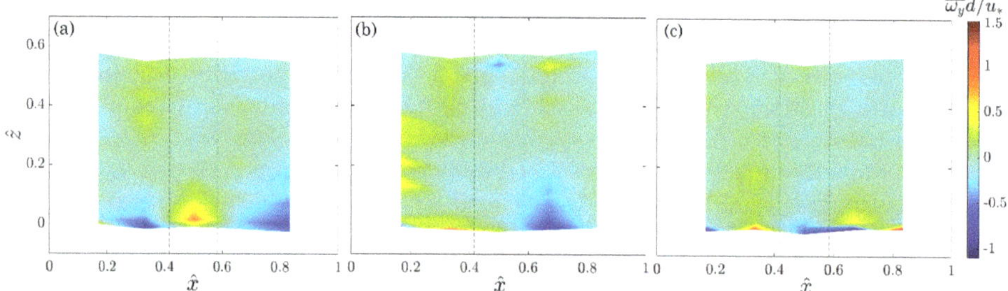

Figure 5. Contours of dimensionless vorticity of the time-averaged flow in the vertical central plane for (**a**) Run 1, (**b**) Run 2, and (**c**) Run 3. The black broken lines indicate the edge of the vegetation stem.

3.2. Anisotropy Invariant Maps

To investigate the anisotropic behavior of the flow through emergent rigid vegetation on rough beds, the AIM was examined along the flume centerline.

Originally introduced by Lumley and Newman [28], this map (also called the Lumley triangle) is a two-dimensional domain based on the invariant properties of the Reynolds stress anisotropy tensor b_{ij}, which can be defined as follows:

$$b_{ij} = \frac{\overline{u'_i u'_j}}{\overline{u'_i u'_i}} - \frac{1}{3}\delta_{ij} \tag{1}$$

where δ_{ij} is the Kronecker delta function ($\delta_{ij}(i = j) = 1$ and $\delta_{ij}(i \neq j) = 0$) and, adopting the Einstein notation, $\overline{u'_i u'_i}$ is twice the TKE. The shape of the AIM is a triangle on a $(III, -II)$ plane (Figure 6), where II is the second invariant of b_{ij} and represents the degree of anisotropy and III is the third invariant of b_{ij} and signifies the nature of anisotropy.

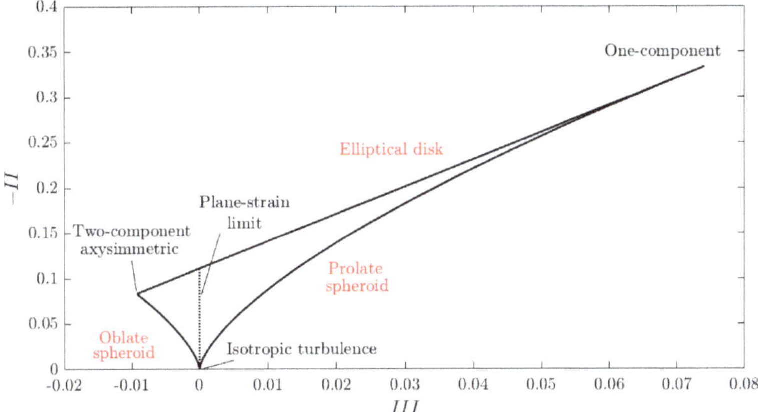

Figure 6. Conceptual diagram of the anisotropy invariant map.

The two invariants can be expressed, respectively, as follows:

$$II = -\frac{b_{ij}b_{ij}}{2} = -\left(\lambda_1^2 + \lambda_1\lambda_2 + \lambda_2^2\right) \tag{2}$$

$$III = \frac{b_{ij}b_{jk}b_{ki}}{3} = -\lambda_1\lambda_2(\lambda_1 + \lambda_2) \tag{3}$$

where λ_1 and λ_2 are the anisotropy eigenvalues.

The AIM is delimited by two curves and an upper line. The left curve is characterized by negative values of the third invariant and can be described as follows: $III = -2(-II/3)^{3/2}$. This curve refers to the pancake-shaped turbulence, since two diagonal components of the Reynolds stress tensor are greater than the third one. The right curve is defined as $III = 2(-II/3)^{3/2}$ and corresponds to the cigar-shaped turbulence, that is, one diagonal component of the Reynolds stress tensor is greater than the other two. Lastly, the function that expresses the upper line is $III = -(9II + 1)/27$; it describes a two-component turbulence. The bottom cusp of the AIM indicates, instead, the 3D isotropic turbulence.

Recently, Dey et al. [29] proposed an additional classification of the turbulence anisotropy, considering the shape of the ellipsoid formed by the Reynolds principal stresses σ_{uu}, σ_{vv}, and σ_{ww} along the x-, y-, and z-axis, respectively. The Reynolds principal stresses are expressed respectively as $\rho\overline{u'u'}$, $\rho\overline{v'v'}$, and $\rho\overline{w'w'}$, where ρ is the mass density of water and v' is the temporal fluctuation of the velocity in the spanwise direction.

Specifically, in the case of $\sigma_{uu} = \sigma_{vv} = \sigma_{ww}$, that is an isotropic turbulence, the stress ellipsoid is a sphere. If $\sigma_{uu} = \sigma_{vv} > \sigma_{ww}$ (on the left curve of the AIM, which is termed as the axisymmetric contraction limit), the stress ellipsoid takes the shape of an oblate spheroid. The two-component axisymmetric limit lies on the left vertex of the AIM, where the conditions $\sigma_{uu} = \sigma_{vv}$ and $\sigma_{ww} = 0$ prevail. In this case, the shape of the stress ellipsoid is a circular disk. On the right curve (the axisymmetric expansion limit), one component of the Reynolds stresses is larger than the other two ($\sigma_{uu} = \sigma_{vv} < \sigma_{ww}$), thus the stress ellipsoid takes the form of a prolate spheroid. At the top boundary, which indicates the two-component limit, the stress ellipsoid is an elliptical disk, since $\sigma_{uu} > \sigma_{vv}$ and $\sigma_{ww} = 0$. Finally, the one-component limit lies on the right vertex, where only one component of Reynolds stress sustains (that is, $\sigma_{uu} > 0$ and $\sigma_{vv} = \sigma_{ww} = 0$ or $\sigma_{ww} > 0$ and $\sigma_{uu} = \sigma_{vv} = 0$). This means that the stress ellipsoid assumes the shape of a straight line.

Figure 7 shows the data plots of $-II$ versus III and the AIMs for Run 1, at the following streamwise relative distances: $x/L_s = 0, 0.17, 0.33, 0.50, 0.67, 0.83, 1.00$. Note that the plots were zoomed in the area of the AIM in which the data were concentrated. In the same way, Figures 8 and 9 depict the AIMs for

Runs 2 and 3, respectively. At a given streamwise distance, each subplot illustrates the evolution of the turbulence anisotropy along the dimensionless vertical distance $\hat{z}$. In all the Runs, moving from the crest level upwards, the data points of the Reynolds stress tensor describe a particular path, with some differences as the bed roughness changes.

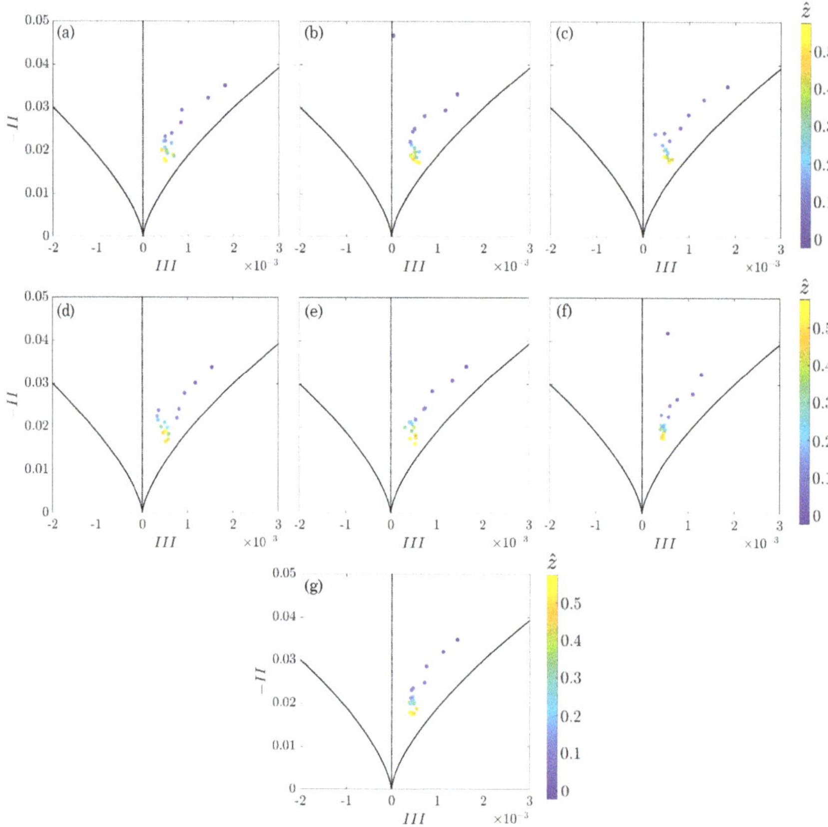

Figure 7. AIMs of Run 1 at (**a**) $x/L_s = 0$, (**b**) $x/L_s = 0.17$, (**c**) $x/L_s = 0.33$, (**d**) $x/L_s = 0.50$ (at the vegetation stem axis), (**e**) $x/L_s = 0.67$, (**f**) $x/L_s = 0.83$, (**g**) $x/L_s = 1.00$.

Looking at Figure 7, near the bed, the turbulence anisotropy is prevalent as the data lie close to the right side of the Lumley triangle. This means that the velocity fluctuation in the vertical direction predominates owing to the bed roughness height, which enhances σ_{ww}. Then, the data plots move toward the line of plane-strain limit, which is characterized by the condition $III = 0$. As the vertical distance increases, the turbulence anisotropy shows a feeble tendency again toward the axisymmetric expansion limit. The described path occurs at each streamwise distance, implying a similar behavior of the Reynolds stress tensor, regardless of the location with respect to the vegetation stem. Therefore, considering the classification based on the ellipsoid shape [29], it is evident that, near the bed, a prolate spheroid axisymmetric turbulence is predominant. Subsequently, as the vertical distance increases, an axisymmetric contraction develops, tending to the 3D isotropic turbulence in the region mainly affected by the vegetation.

As regards Run 2, although near the crest level, the position of the data points in the AIMs does not vary from that of Run 1; moving toward the free surface, their path slightly changes. In fact, it is evident that the data plots tend to move toward the line of plane-strain limit, but they rapidly turn back to the right side of the Lumley triangle. This implies that one component of the Reynold stresses prevails on the others for almost the entire investigated flow depth. Thus, in Run 2, the ellipsoid is basically a prolate spheroid along $\hat{z}$. The same trend is visible at the different streamwise distances.

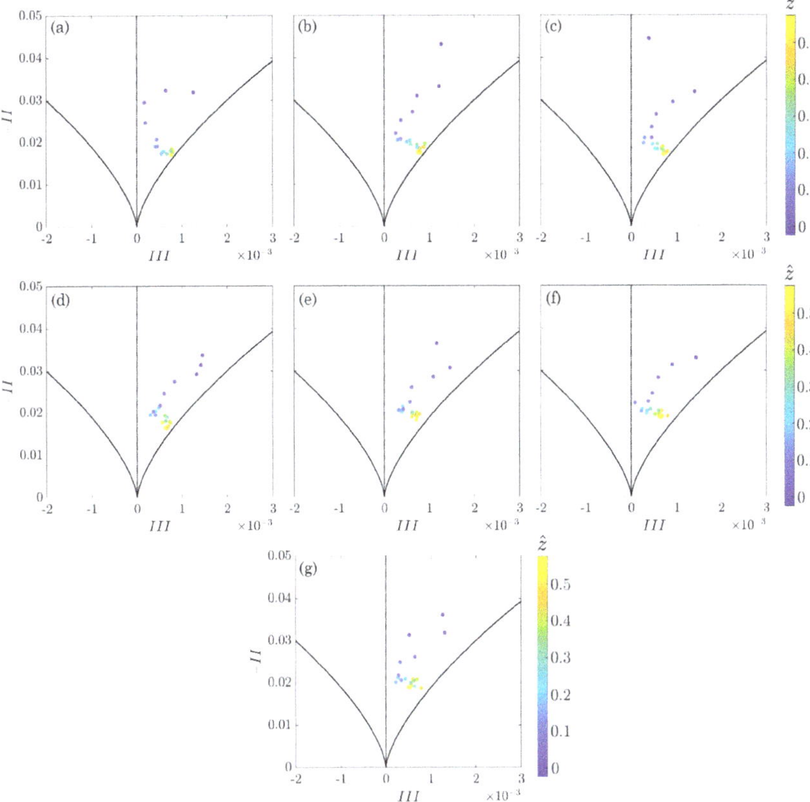

Figure 8. AIMs of Run 2 at (**a**) $x/L_s = 0$, (**b**) $x/L_s = 0.17$, (**c**) $x/L_s = 0.33$, (**d**) $x/L_s = 0.50$ (at the vegetation stem axis), (**e**) $x/L_s = 0.67$, (**f**) $x/L_s = 0.83$, (**g**) $x/L_s = 1.00$.

Akin to both Runs 1 and 2, the data plots of Run 3 start from the axisymmetric expansion limit (that corresponds to the right-curved side of the triangle). Increasing the vertical distance, they rapidly move toward the line of plane-strain limit. Then, the turbulence anisotropy tends to the isotropic state (the data move toward the bottom cusp of the AIM). This is due to the bed roughness influence on turbulence anisotropy, which vanishes moving toward the free surface. Thus, initially the ellipsoid shape is a prolate spheroid. As the vertical distance increases, an axisymmetric contraction develops, tending to the 3D isotropic state and, as a result, the stress ellipsoid becomes a sphere.

In order to highlight the effects induced by the presence of vegetation, Figure 10 shows the AIMs for the undisturbed flow conditions detected 50 cm upstream to the vegetation array in all the three Runs. It is revealed that, without the influence of vegetation, the turbulence anisotropy tends to the plane-strain limit in all the Runs, independently from the bed roughness, which, however, is the main cause of a prolate spheroid axisymmetric turbulence near the bed surface. This latter is more pronounced in Run 3, owing to a higher roughness height than in the other two Runs.

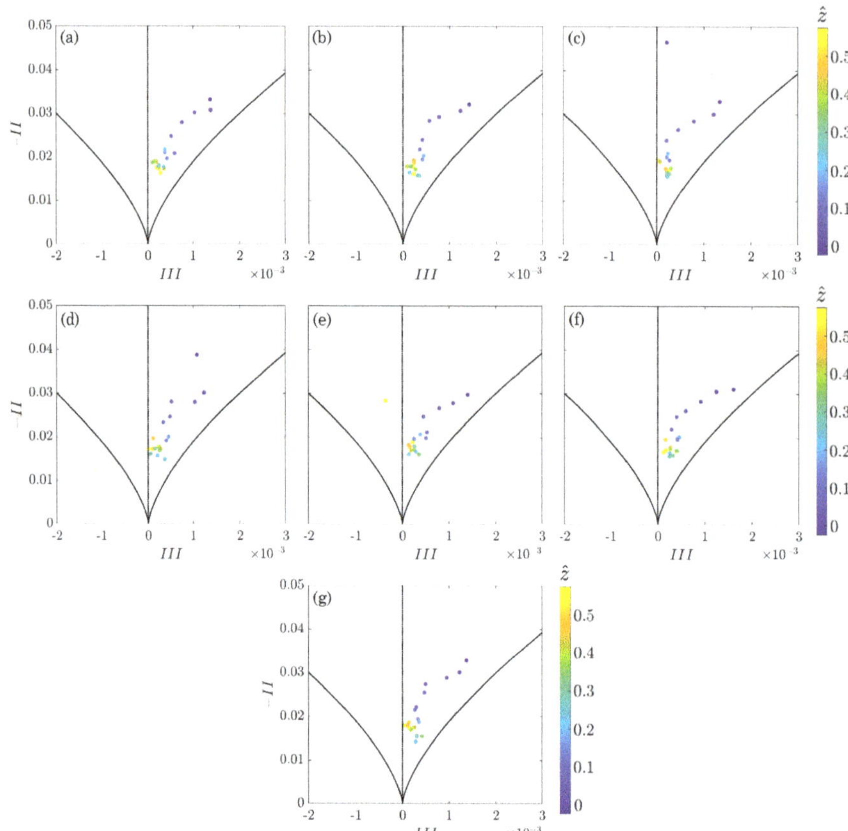

Figure 9. AIMs of Run 3 at (**a**) $x/L_S = 0$, (**b**) $x/L_S = 0.17$, (**c**) $x/L_S = 0.33$, (**d**) $x/L_S = 0.50$ (at the vegetation stem axis), (**e**) $x/L_S = 0.67$, (**f**) $x/L_S = 0.83$, (**g**) $x/L_S = 1.00$.

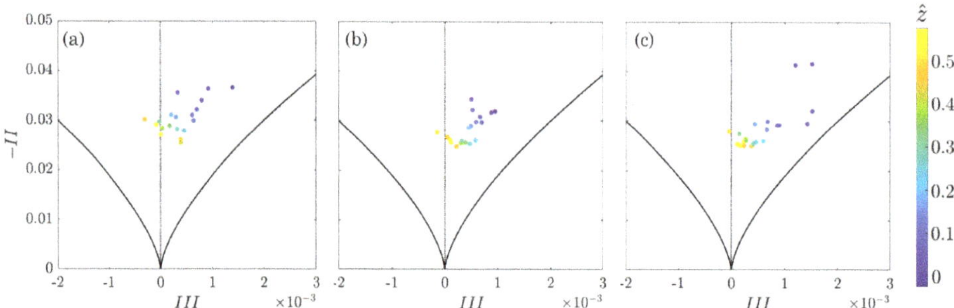

Figure 10. AIMs of the undisturbed flow condition for (**a**) Run 1, (**b**) Run 2, and (**c**) Run 3.

3.3. Anisotropic Invariant Function

Choi and Lumley [30] introduced a function, called the anisotropic invariant function F, with the aim of providing an insight into the turbulence anisotropy from the two-component limit to the isotropic limit [31]. The function can be calculated as follows:

$$F = 1 + 9II + 27III \tag{4}$$

The main peculiarity of the anisotropic invariant function is that it vanishes when the turbulence anisotropy prevails ($F = 0$), whereas it reaches unity ($F = 1$) when the turbulence reaches the three-dimensional isotropic state.

The contours of the anisotropic invariant function on the vertical central plane in the test section are shown in Figure 11 for all the Runs. The streamwise variation of the anisotropic invariant function is quite uniform, regardless of the location of the vegetation stem. However, it is possible to note that the topographical configuration of the bed surface has a strong impact on the turbulence characteristics of the flow. In fact, on the uphill stretches F is almost null, indicating a strong two-dimensional turbulence, since one velocity component is limited by the bed. Then, the anisotropic invariant function becomes greater than 0 on the downhill stretches, where the turbulence can develop in the three directions.

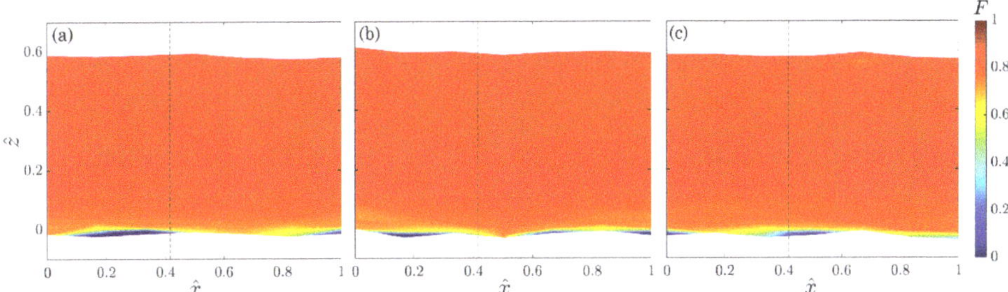

Figure 11. Contours of the anisotropic invariant function F in the test section for (**a**) Run 1, (**b**) Run 2, and (**c**) Run 3. The black broken lines indicate the edge of the vegetation stem.

Moving toward the free surface, the anisotropic invariant function gradually increases, reaching approximately $F = 0.9$. This confirms that the bed roughness influence on turbulence anisotropy vanishes moving toward the free surface.

3.4. New Research Prospects

The experimental results obtained in this study represent a new dataset that may be used for the calibration of advanced numerical models, which are usually based on isotropic turbulence hypothesis. In fact, as it was demonstrated, vegetation and bed roughness can hinder flow by acting as an obstruction, generating turbulence, and affecting the entire flow velocity distribution [32], modifying the turbulence behavior from isotropic to anisotropic moving toward the bed surface.

The modelling of such vegetated flows on rough beds clearly gets complicated if natural and complex channel cross-sections with different shapes are considered. Recent researches showed that different rectangular and trapezoidal shapes as well as the corner angles can exert a strong impact on the flow velocity distribution and its induced secondary flow [33–35]; hence, they may affect the sediment transport process [36]. Therefore, new analytical models of sidewall turbulence effect on streamwise velocity profile have been recently proposed (e.g., [35,37]) for uneven narrow and wide channels. These analytical models could be improved by considering extra turbulence zones represented by both vegetation and rough beds.

However, flow structures become even more complex when vegetation appears in channels with bedforms [38]. How the existence of vegetation changes flow pattern over gravel bedforms still remains poorly understood and could be considered as another potential development of the present research work.

4. Conclusions

The aim of the present work was the study of the turbulence anisotropy in the free stream region of turbulent flows through rigid emergent vegetation on rough beds using AIMs. Three experimental runs were performed in a uniformly distributed vegetated channel with three different bed sediments. The principal findings are summarized below.

1. Two different zones were characterized along the flume centerline: a convergent flow zone between two stems and toward the flume centerline; a retarded flow zone owing to a divergent flow downstream of the stems. Owing to the bed roughness, a near-bed flow heterogeneity was found, activating the fluid streaks to have motions alternatively in both clockwise and counterclockwise directions. Here, the Reynolds shear stresses increase with a high gradient, but at the crest level they become negligible and the viscous stresses reach their maximum values. However, the quite uniform streamwise distribution of both $\hat{\tau}_{uw}$ and $\hat{\tau}_v$ reveal that they were not influenced by the position of vegetation stems.

2. The analysis of the AIMs revealed that the combined effect of vegetation and bed roughness causes the evolution of the turbulence from the quasi-three-dimensional isotropy (the stress ellipsoid is like a sphere) to a prolate spheroid axisymmetric turbulence. This kind of turbulence anisotropy is kept also near the bed surface. This particular pattern is also confirmed by the contours of the anisotropic invariant function.

3. The topographical configuration of the bed surface has a strong impact on the turbulent characteristics of the flow. In fact, on the uphill stretches, the anisotropic invariant function indicates a strong two-dimensional turbulence, since one velocity component is limited by the bed surface. Instead, on the downhill stretches, the anisotropic invariant function reveals that the turbulence can develop in the three directions.

Author Contributions: Conceptualization, N.P., F.C., A.D., R.G.; methodology, N.P., F.C., A.D., R.G.; formal analysis, N.P., F.C.; data curation, F.C.; writing—original draft preparation, N.P.; writing—review and editing, N.P., F.C., A.D., R.G.; supervision, R.G.; funding acquisition, R.G. All authors have read and agreed to the published version of the manuscript.

Funding: This research was funded by the *"PreFluSed—Prevenzione del rischio di alluvioni in un bacino Fluviale calabrese in presenza di trasporto di Sedimenti"* Project (*Ministero dell'Ambiente e della Tutela del Territorio e del Mare, Direzione Generale per la Salvaguardia del Territorio e delle Acque*, Italy).

Acknowledgments: The authors would like to thank Davide Garigliano for his valuable work during the performance of the experimental runs and the anonymous referees for their suggestions and comments.

Conflicts of Interest: The authors declare no conflict of interest.

References

1. White, B.L.; Nepf, H.M. A vortex-based model of velocity and shear stress in a partially vegetated shallow channel. *Water Resour. Res.* **2008**, *44*, 1–15. [CrossRef]
2. Ben Meftah, M.; De Serio, F.; Mossa, M. Hydrodynamic behavior in the outer shear layer of partly obstructed open channels. *Phys. Fluids* **2014**, *26*, 065102. [CrossRef]
3. Caroppi, G.; Gualtieri, P.; Fontana, N.; Giugni, M. Vegetated channel flows: Turbulence anisotropy at flow–rigid canopy interface. *Geosciences* **2018**, *8*, 259. [CrossRef]
4. Rowiński, P.M.; Västilä, K.; Aberle, J.; Järvelä, J.; Kalinowska, M.B. How vegetation can aid in coping with river management challenges: A brief review. *Ecohydrol. Hydrobiol.* **2018**, *18*, 345–354. [CrossRef]
5. Caroppi, G.; Västilä, K.; Järvelä, J.; Rowiński, P.M.; Giugni, M. Turbulence at water-vegetation interface in open channel flow: Experiments with natural-like plants. *Adv. Water Resour.* **2019**, *127*, 180–191. [CrossRef]
6. Ricardo, A.M.; Franca, M.J.; Ferreira, R.M. Turbulent flows within random arrays of rigid and emergent cylinders with varying distribution. *J. Hydraul. Eng.* **2016**, *142*, 04016022. [CrossRef]
7. Schoelynck, J.; De Groote, T.; Bal, K.; Vandenbruwaene, W.; Meire, P.; Temmerman, S. Self-organised patchiness and scale dependent bio-geomorphic feedbacks in aquatic river vegetation. *Ecography* **2012**, *35*, 760–768. [CrossRef]
8. Bearman, P.W.; Zdravkovich, M.M. Flow around a circular cylinder near a plane boundary. *J. Fluid Mech.* **1978**, *89*, 33–47. [CrossRef]

9. Nezu, I.; Sanjou, M. Turburence structure and coherent motion in vegetated canopy open-channel flows. *J. Hydro-Environ. Res.* **2008**, *2*, 62–90. [CrossRef]

10. Yang, W.; Choi, S.U. A two-layer approach for depth-limited open-channel flows with submerged vegetation. *J. Hydraul. Res.* **2010**, *48*, 466–475. [CrossRef]

11. Shimizu, Y.; Tsujimoto, T. Numerical analysis of turbulent open-channel flow over a vegetation layer using a k-ε turbulence model. *J. Hydrosci. Hydraul. Eng.* **1994**, *11*, 57–67.

12. Nepf, H. Drag, turbulence, and diffusion in flow through emergent vegetation. *Water Resour. Res.* **1999**, *35*, 479–489. [CrossRef]

13. Righetti, M.; Armanini, A. Flow resistance in open channel flows with sparsely distributed bushes. *J. Hydrol.* **2002**, *269*, 55–64. [CrossRef]

14. Choi, S.U.; Kang, H. Numerical investigations of mean flow and turbulence structures of partly-vegetated open-channel flows using the Reynolds stress model. *J. Hydraul. Res.* **2006**, *44*, 203–217. [CrossRef]

15. Poggi, D.; Krug, C.; Katul, G.G. Hydraulic resistance of submerged rigid vegetation derived from first-order closure models. *Water Resour. Res.* **2009**, *45*, W10442. [CrossRef]

16. Gualtieri, P.; De Felice, S.; Pasquino, V.; Doria, G. Use of conventional flow resistance equations and a model for the Nikuradse roughness in vegetated flows at high submergence. *J. Hydrol. Hydromech.* **2018**, *66*, 107–120. [CrossRef]

17. Maji, S.; Hanmaiahgari, P.R.; Balachandar, R.; Pu, J.H.; Ricardo, A.M.; Ferreira, R.M. A Review on Hydrodynamics of Free Surface Flows in Emergent Vegetated Channels. *Water* **2020**, *12*, 1218. [CrossRef]

18. Penna, N.; Coscarella, F.; D'Ippolito, A.; Gaudio, R. Bed roughness effects on the turbulence characteristics of flows through emergent rigid vegetation. *Water* **2020**, *12*, 2401. [CrossRef]

19. Emory, M.; Iaccarino, G. Visualizing turbulence anisotropy in the spatial domain with componentality contours. *Cent. Turbul. Res. Annu. Res. Briefs* **2014**, 123–138.

20. Sarkar, S.; Ali, S.Z.; Dey, S. Turbulence in Wall-Wake Flow Downstream of an Isolated Dunal Bedform. *Water* **2019**, *11*, 1975. [CrossRef]

21. Nepf, H.M. Flow and transport in regions with aquatic vegetation. *Annu. Rev. Fluid Mech.* **2012**, *44*, 123–142. [CrossRef]

22. Julien, P.Y. *Erosion and Sedimentation*; Cambridge University Press: Cambridge, UK, 1998.

23. Goring, D.G.; Nikora, V.I. Despiking acoustic Doppler velocimeter data. *J. Hydraul. Eng.* **2002**, *128*, 117–126. [CrossRef]

24. De Serio, F.; Ben Meftah, M.; Mossa, M.; Termini, D. Experimental investigation on dispersion mechanisms in rigid and flexible vegetated beds. *Adv. Water Resour.* **2018**, *120*, 98–113. [CrossRef]

25. Padhi, E.; Penna, N.; Dey, S.; Gaudio, R. Near-bed turbulence structures in water-worked and screeded gravel-bed flows. *Phys. Fluids* **2019**, *31*, 045107. [CrossRef]

26. Penna, N.; Coscarella, F.; Gaudio, R. Turbulent Flow Field around Horizontal Cylinders with Scour Hole. *Water* **2020**, *12*, 143. [CrossRef]

27. Ricardo, A.M.; Koll, K.; Franca, M.J.; Schleiss, A.J.; Ferreira, R.M.L. The terms of turbulent kinetic energy budget within random arrays of emergent cylinders. *Water Resour. Res.* **2014**, *50*, 4131–4148. [CrossRef]

28. Lumley, J.L.; Newman, G.R. The return to isotropy of homogeneous turbulence. *J. Fluid Mech.* **1977**, *82*, 161–178. [CrossRef]

29. Dey, S.; Ravi Kishore, G.; Castro-Orgaz, O.; Ali, S.Z. Turbulent length scales and anisotropy in submerged turbulent plane offset jets. *J. Hydraul. Eng.* **2019**, *145*, 04018085. [CrossRef]

30. Choi, K.S.; Lumley, J.L. The return to isotropy of homogeneous turbulence. *J. Fluid Mech.* **2001**, *436*, 59–84. [CrossRef]

31. Penna, N.; Padhi, E.; Dey, S.; Gaudio, R. Structure functions and invariants of the anisotropic Reynolds stress tensor in turbulent flows on water-worked gravel beds. *Phys. Fluids* **2020**, *32*, 055106. [CrossRef]

32. Pu, J.H.; Hussain, A.; Guo, Y.K.; Vardakastanis, N.; Hanmaiahgari, P.R.; Lam, D. Submerged flexible vegetation impact on open channel flow velocity distribution: An analytical modelling study on drag and friction. *Water Sci. Eng.* **2019**, *12*, 121–128. [CrossRef]

33. Lucas, J.; Lutz, N.; Lais, A.; Hager, W.H.; Boes, R.M. Side-channel flow: Physical model studies. *J. Hydraul. Eng.* **2017**, *05015003*, 1–11. [CrossRef]

34. Vidal, A.; Nagib, H.M.; Schlatter, P.; Vinuesa, R. Secondary flow in spanwise periodic in-phase sinusoidal channels. *J. Fluid Mech.* **2018**, *851*, 288–316. [CrossRef]

35. Pu, J.H.; Pandey, M.; Hanmaiahgari, P.R. Analytical modelling of sidewall turbulence effect on streamwise velocity profile using 2D approach: A comparison of rectangular and trapezoidal open channel flows. *J. Hydro-Environ. Res.* **2020**. [CrossRef]
36. Pu, J.H.; Lim, S.Y. Efficient numerical computation and experimental study of temporally long equilibrium scour development around abutment. *Environ. Fluid Mech.* **2014**, *14*, 69–86. [CrossRef]
37. Pu, J.H. Turbulent rectangular compound open channel flow study using multi-zonal approach. *Environ. Fluid Mech.* **2019**, *19*, 785–800. [CrossRef]
38. Afzalimehr, H.; Maddahi, M.R.; Sui, J.; Rahimpour, M. Impacts of vegetation over bedforms on flow characteristics in gravel-bed rivers. *J. Hydrodyn.* **2019**, *31*, 986–998. [CrossRef]

Article

Turbulence in Wall-Wake Flow Downstream of an Isolated Dunal Bedform

Sankar Sarkar [1], Sk Zeeshan Ali [2] and Subhasish Dey [2,*]

[1] Physics and Applied Mathematics Unit, Indian Statistical Institute, Kolkata, West Bengal 700108, India; sankar_s@isical.ac.in

[2] Department of Civil Engineering, Indian Institute of Technology Kharagpur, West Bengal 721302, India; skzeeshanali@iitkgp.ac.in

* Correspondence: sdey@iitkgp.ac.in; Tel.: +91-943-471-3850

Received: 1 September 2019; Accepted: 19 September 2019; Published: 22 September 2019

Abstract: This study examines the turbulence in wall-wake flow downstream of an isolated dunal bedform. The streamwise flow velocity and Reynolds shear stress profiles at the upstream and various streamwise distances downstream of the dune were obtained. The results reveal that in the wall-wake flow, the third-order moments change their signs below the dune crest, whereas their signs remain unaltered above the crest. The near-wake flow is featured by sweep events, whereas the far-wake flow is controlled by the ejection events. Downstream of the dune, the turbulent kinetic energy production and dissipation rates, in the near-bed flow zone, are positive. However, they reduce as the vertical distance increases up to the lower-half of the dune height and beyond that, they increase with an increase in vertical distance, attaining their peaks at the crest. The turbulent kinetic energy diffusion and pressure energy diffusion rates, in the near-bed flow zone, are negative, whereas they attain their positive peaks at the crest. The anisotropy invariant maps indicate that the data plots in the wall-wake flow form a looping trend. Below the crest, the turbulence has an affinity to a two-dimensional isotropy, whereas above the crest, the anisotropy tends to reduce to a quasi-three-dimensional isotropy.

Keywords: hydraulics; turbulent flow; wall-wake flow; dunal bedform

---

1. Introduction

Turbulent flow over dunal bedforms fascinates researchers. The topic is important not only from the viewpoint of intrinsic scientific reasons, but also owing to its far-reaching applications in engineering. In addition to its practical applications, it allows a significant theoretical understanding of wake flows. Despite impressive advances over the past years, an inclusive picture of the flow and turbulence characteristics over a dunal bedform remains far from complete [1]. The dunes are created by an interaction between the flow and bed sediment particles. Dunes are kind of bedforms that are found when the flow variables, such as flow velocity and bed shear stress over a sediment bed surpass their threshold values.

Over the decades, a large corpus of experimental and numerical studies has been reported to grasp the flow features over dunal bedforms. Researchers studied the velocity field over dunes to acquire an insight into the physical features, including the reattachment point, wake region and internal boundary layer [2,3]. The experimental observations of flow over a series of two- and three-dimensional dunes revealed that the two-dimensional dunes induce stronger turbulence compared to their three-dimensional counterparts [4,5]. However, the flow characteristics over a natural dune were found to be quite different from those over an artificial dune [6]. Best [7] found that over the dune crests, the ejections dominate the instantaneous flow field.

In a natural streamflow, an isolated dunal bedform acts as a bluff-body, producing wall-wake flow at its downstream. The wake flow downstream of an isolated dunal bedform persists up to a certain stretch until the local wake flow diffuses to and becomes the part of the undisturbed upstream flow. Figure 1 presents a conceptual representation of flow past an isolated dunal bedform in xz plane. Here, x is the streamwise distance measured from a convenient point O and z is the vertical distance from the bed. The dune length L_d comprises the stoss-side length L_s and the leeside length L_l ($L_d = L_s + L_l$). The dune height H_d is the vertical distance of the dune crest from the bed. Downstream of the dune, a flow reversal takes place, called the *near-wake flow*. Afterward, the flow is called the *far-wake flow*. In Figure 1, the lower dashed line denotes the locus of $\bar{u}(z) = 0$, whereas the upper dashed line signifies the boundary layer ($\bar{u} = \bar{u}_0$) in the wall-wake flow. Here, $\bar{u}(z)$ is the time-averaged streamwise flow velocity in the wake flow and $\bar{u}_0(z)$ is the time-averaged streamwise flow velocity in the undisturbed upstream flow. In the far downstream of the dunal bedform, the flow achieves the undisturbed upstream state, called the *fully recovered* open-channel flow.

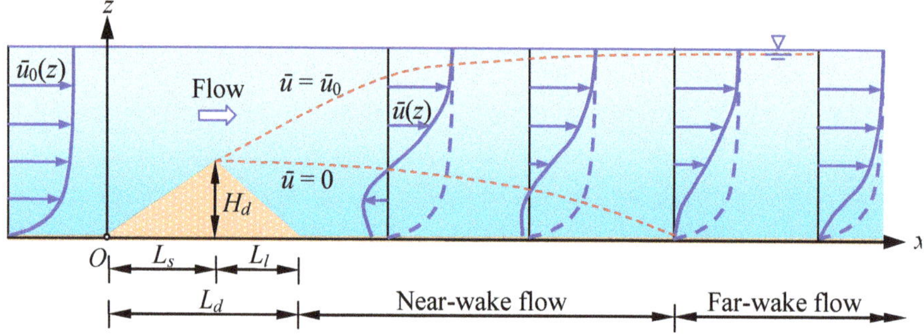

Figure 1. Conceptual sketch of flow over an isolated dunal bedform.

In this context, it is pertinent to mention that for a shear-free flow, Schlichting [8] pioneered the similarity theory of the velocity defect profile in the free-wake flow downstream of a circular cylinder. The wall-wake flow downstream of an isolated dunal bedform in an approach wall-shear flow, being different from a free-wake flow, is rather intricate. The turbulence characteristics and the vortex shedding downstream of bed-mounted bluff-bodies in both near- and far-wake flows were studied by various researchers. Some of these bluff-bodies include plate [9,10], hemisphere [11], sphere [12,13], circular cylinder [14–21] and pebble cluster [22].

It is worth noting that most of the former studies were dedicated to understanding the flow features over a continuous train of dunes. In fact, little is known about the flow and turbulence characteristics over an isolated dunal bedform. This study specifically puts into focus the flow and turbulence characteristics downstream of an isolated two-dimensional dunal bedform over a rough bed in order to advance the present state-of-the-art. In addition to time-averaged streamwise flow velocity, the salient features of turbulence, including the Reynolds shear stress, turbulent bursting, turbulent kinetic energy budget and Reynolds stress anisotropy, are greatly discussed. It may be noted that the preliminary studies of flow and turbulence characteristics downstream of an isolated dunal bedform have been recently presented elsewhere [23,24].

2. Experimental Design

Experiments were performed in a re-circulatory flume, having a rectangular cross-section, at the Fluvial Mechanics Laboratory in the Indian Statistical Institute, Kolkata, India. The length, width and height of the flume were 20 m, 0.5 m and 0.5 m, respectively. The inflow discharge, supplied by a centrifugal pump, was measured by an electromagnetic gadget. The transparent sidewalls of the flume provided visual access to the flow. The flume bed, having a streamwise bed slope of 3×10^{-4}, was prepared by gluing uniform gravels of median size d_{50} = 2.49 mm. In the experiments, two types of isolated two-dimensional dunal bedforms, classified as Runs 1 and 2, respectively (Figure 2), were mounted on the flume bed at a distance of 7 m from the inlet. In Runs 1 and 2, the dune heights H_d were 0.09 m and 0.03 m, whereas the dune lengths L_d were 0.4 m (L_S = 0.24 m and L_l = 0.16 m) and 0.3 m (L_S = 0.24 m and L_l = 0.06 m), respectively. In both the runs, the same approach uniform flow condition was maintained. The approach flow depth h and depth-averaged approach flow velocity $\bar{U}_0$ were maintained as $h \approx 0.3$ m and $\bar{U}_0 \approx 0.44$ m s^{-1}. The flow depth and the free surface profile were measured by a Vernier point gauge, having a precision of ±0.1 mm. The approach shear velocity u_* [= $(\tau_0/\rho)^{0.5}$], obtained from the streamwise bed slope, was 0.03 m s^{-1}. Here, τ_0 is the bed shear stress and ρ is the mass density of fluid. However, the values of u_* in both Runs 1 and 2, determined from the Reynolds shear stress profiles, were 0.027 m s^{-1} and 0.025 m s^{-1}, respectively. It is worth noting that to find the u_* from the Reynolds shear stress profiles, the profiles were extrapolated up to the bed. In both the runs, the flow Reynolds number was 528,000, whereas the flow Froude number was 0.256 (subcritical). The shear Reynolds number R_* (= $d_{50}u_*/\nu$, where ν is the coefficient of kinematic viscosity of fluid) was preserved to be 74.7 (> 70), setting a hydraulically rough flow regime.

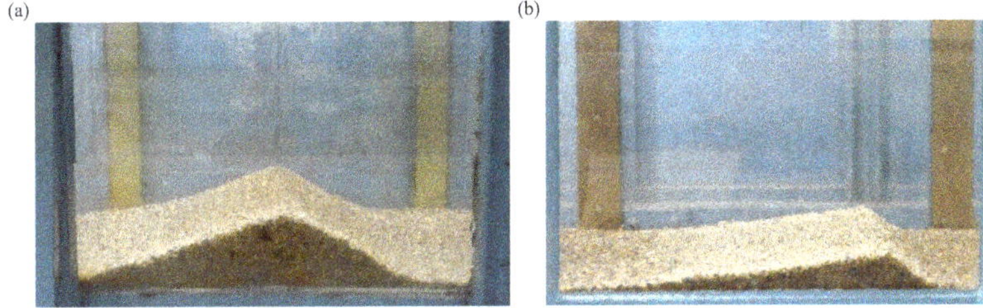

Figure 2. Photographs of isolated dunal bedforms in (**a**) Run 1 and (**b**) Run 2. Flow direction is from left to right.

A 5 cm down-looking *Vecrtino* probe (acoustic Doppler velocimetry), also called *Vectrino plus*, was used to capture the instantaneous three-dimensional flow velocity components along the flume centreline at various relative streamwise distances x/L_d = −0.5, −0.25, 0, 0.1, 0.2, 0.3, 0.4, 0.5, 0.6, 0.7, 0.8, 0.9, 1, 1.1, 1.3, 1.7, 2.1, 2.5 and 3.3. The Vecrtino system, having a flexible sampling volume of 6 mm diameter and 1 to 4 mm height, was operated with 10 MHz acoustic frequency and 100 Hz sampling rate. The velocity components (u, v, w) correspond to (x, y, z), where y is the spanwise direction. It may be noted that up to the dune crest, the lowest sampling height was set as 1 mm, whereas beyond the crest, it was 2.5 mm. The closest measuring location of the data points was 2 mm. A sampling duration of 300 s was found to be adequate to obtain the time-independent flow velocity and turbulence quantities. The minimum signal-to-noise ratio was maintained as 18, whereas the minimum threshold of signal correlation was maintained as 70%. The measured data were filtered whenever required applying the *acceleration thresholding method* [25]. This method could separate and substitute the unwanted data spikes in two phases. The threshold values of 1 to 1.5 for decontaminating the measured data were ascertained by satisfying Kolmogorov '−5/3' scaling law in the inertial subrange for the spectral density function $S_{df}(k_w)$ of streamwise velocity fluctuations u'. Here, k_w is the wavenumber (= $2\pi f/\bar{u}$) and f is the frequency. Figure 3a,b illustrates the data plots of $S_{df}(k_w)$ for velocity fluctuations (u', v', w') in (x, y, z) before and after decontaminating the data in Run 1, respectively, at a relative streamwise distance x/L_d = 0.7 and a relative vertical distance z/L_d = 0.13. The $S_{df}(k_w)$ curves of decontaminated signals compare well with Kolmogorov '−5/3' scaling law in the inertial subrange for $k_w \geq 30$ rad s^{-1}. In addition, it appears that the discrete spectral peaks are prominent for $k_w < 30$ rad s^{-1}. This indicates that the signals corresponding to $k_w < 30$ rad s^{-1} contained large-scale turbulent structures, while those for $k_w \geq 30$ rad s^{-1} confirmed a pure turbulence. Therefore, a high-pass filter with a cut-off wavenumber of 30 rad s^{-1} was used to filter the data.

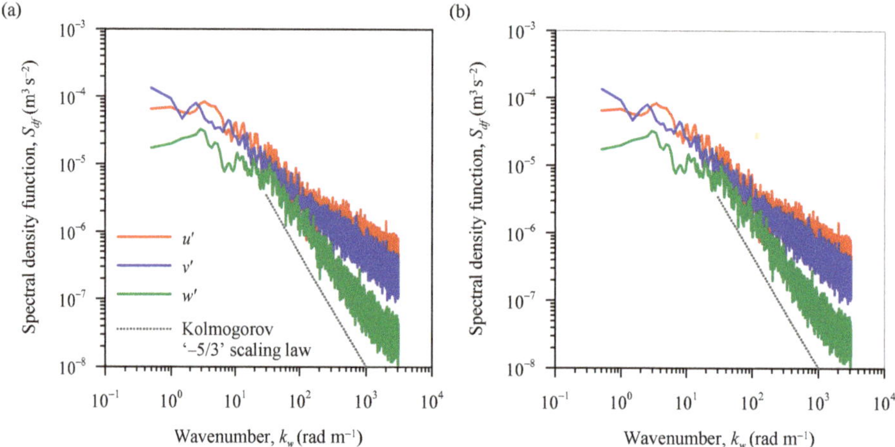

Figure 3. Spectral density function $S_{df}(k_w)$ versus wavenumber k_w (**a**) before and (**b**) after decontaminating the data in Run 1 at a relative streamwise distance $x/L_d = 0.7$ and a relative vertical distance $z/L_d = 0.13$.

In order to find the uncertainty of Vectrino data, 15 samples were collected at a sampling rate of 100 Hz for a duration of 300 s at a vertical distance $z = 5$ mm. Table 1 summarizes the results of uncertainty estimations of the time-averaged velocity components $(\bar{u}, \bar{v}, \bar{w})$ and the turbulence intensities $[(\overline{u'u'})^{0.5}, (\overline{v'v'})^{0.5}, (\overline{w'w'})^{0.5}]$ in (x, y, z) and the Reynolds shear stress τ per unit mass density of fluid $(= -\overline{u'w'})$. It is pertinent to mention that to avoid bias and random errors, the samplings were done every time after resuming the experiments. The errors for the time-averaged velocity components, turbulence intensities and Reynolds shear stress were within ±4%, ±7% and ±8%, respectively. This confirmed the appropriateness of the data sampling with 100 Hz sampling rate. Further, it was necessary to ascertain the fully-developed undisturbed approach velocity profiles for both the Runs. Figure 4 shows the vertical profiles of nondimensional streamwise flow velocity $\bar{u}^+$ ($= \bar{u}/u_*$) at the upstream of isolated dunal bedforms for both Runs 1 and 2. The data plots compare well with the classical logarithmic law $\bar{u}/u_* = \kappa^{-1}\ln(z/d_{50}) + 8.5$ for a hydraulically rough flow regime. Here, κ is the von Kármán constant ($= 0.41$). This confirmed the acceptability of the fully-developed undisturbed approach flow velocity profiles for a hydraulically rough flow regime.

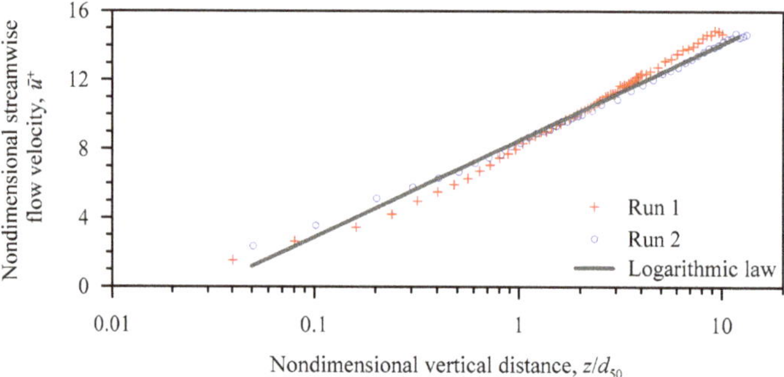

Figure 4. Vertical profiles of nondimensional streamwise flow velocity $\bar{u}^+$ at the upstream of isolated dunal bedforms for Runs 1 and 2.

Table 1. Uncertainty estimation for Vectrino.

$\tilde{U}$ (m s^{-1})	$\bar{v}$ (m s^{-1})	$\bar{w}$ (m s^{-1})	$\overline{(u'u')}^{0.5}$ (m s^{-1})	$\overline{(v'v')}^{0.5}$ (m s^{-1})	$\overline{(w'w')}^{0.5}$ (m s^{-1})	τ (m^2 s^{-2})
2.94×10^{-3} *	2.33×10^{-3}	1.75×10^{-3}	2.18×10^{-3}	1.34×10^{-3}	1.07×10^{-3}	4.37×10^{-5}
$(\pm 2.93 \times 10^{-2}$ †)	$(\pm 3.02 \times 10^{-2})$	$(\pm 3.95 \times 10^{-2})$	$(\pm 5.87 \times 10^{-2})$	$(\pm 6.72 \times 10^{-2})$	$(\pm 6.89 \times 10^{-2})$	$(\pm 7.48 \times 10^{-4})$

* Standard deviation. † Average of maximum (negative and positive) percentage error.

3. Time-Averaged Flow

3.1. Streamwise Flow Velocity

Figure 5 shows the vertical profiles of nondimensional streamwise flow velocity $\bar{u}^+$ at upstream and various downstream relative streamwise distances x/L_d in Runs 1 and 2. Immediate downstream of the dune ($x/L_d = 1$), the wall-shear flow separates from the dune crest, giving rise to a flow reversal owing to negative streamwise flow velocity. The near-wake flow zone extends up to $x/L_d \approx 1.7$. As the flow reaches further downstream, the flow reversal disappears. In addition, the streamwise flow velocity, having a velocity defect, starts to recover the undisturbed upstream velocity profile in the far-wake zone ($x/L_d = 2.1$ to 2.5). At $x/L_d \approx 3.3$, the velocity profile appears to follow the undisturbed upstream velocity profile. It is also evident that above the relative vertical distance $z/H_d = 1.5$, the values of $\bar{u}^+$ remain almost the same irrespective of x/L_d. However, the extents of the near- and far-wake flow zones in Runs 1 and 2 are different because of the effects of dune dimensions. It is worth mentioning that in wall-wake flows downstream of a sphere and a horizontal cylinder, the velocity profiles appear to follow their corresponding undisturbed upstream velocity profile at streamwise distances equaling roughly 8.5 and 7 times the diameter of sphere and cylinder, respectively [12,21].

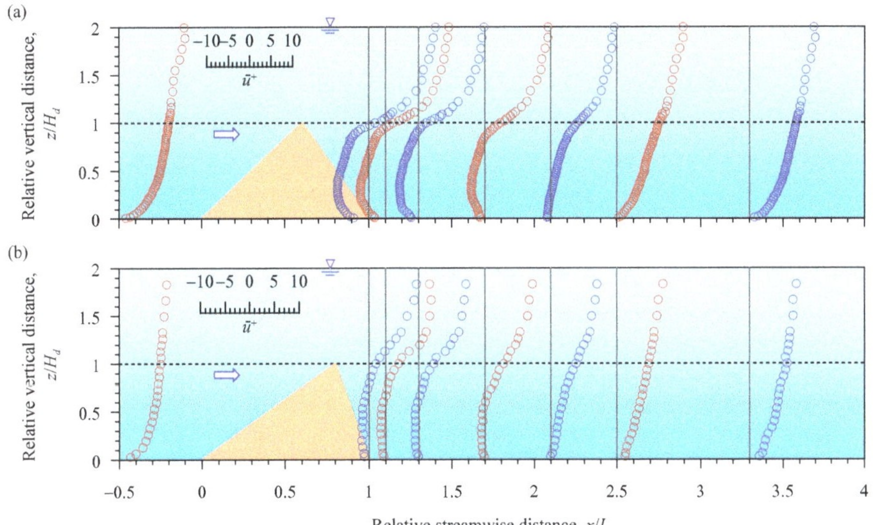

Figure 5. Vertical profiles of nondimensional streamwise flow velocity $\bar{u}^+$ at the upstream and various downstream relative streamwise distances x/L_d of isolated dunal bedforms for (**a**) Run 1 and (**b**) Run 2.

3.2. Reynolds Shear Stress

Figure 6 presents the vertical profiles of nondimensional Reynolds shear stress τ^+ ($= \tau/u_*^2$) at the upstream and various downstream relative streamwise distances x/L_d in Runs 1 and 2. Upstream of the dune ($x/L_d = -0.5$), the τ^+ profile follows a linear law. The τ^+ is approximately unity at the relative vertical distance $z/H_d = 0$ and then, it reduces with an increase in relative vertical distance to become zero at the free surface (if the profiles would be extended up to the free surface). Immediate downstream of the dune ($x/L_d = 1$), the τ^+ is negative in the near-bed flow zone. Thereafter, it increases with an increase in z/H_d, attaining a positive peak at the dune crest ($z/H_d = 1$). Above the crest, the τ^+ decreases with an increase in z/H_d and attains almost similar pattern to the upstream profile for $z/H_d > 1.5$. It appears that for a given z/H_d, the τ^+ decreases with an increase in x/L_d. In particular, at $x/L_d \approx 3.3$, the τ^+ profile becomes almost similar to the upstream profile at $x/L_d = -0.5$. It may be noted that for $z/H_d > 1.75$, the values of τ^+ at various x/L_d are nearly similar. Therefore, it may be concluded that the Reynolds shear stress in the wall-wake flow is influenced by the dune up to a vertical distance of approximately 1.75 times the dune height and a streamwise distance of approximately 2.5 times the dune length.

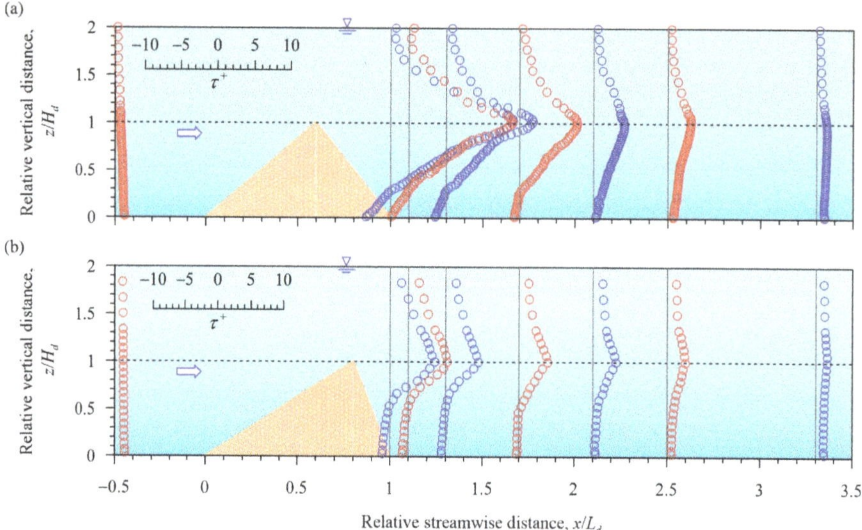

Figure 6. Vertical profiles of nondimensional Reynolds shear stress τ^+ at the upstream and various downstream relative streamwise distances x/L_d of isolated dunal bedforms for (**a**) Run 1 and (**b**) Run 2.

4. Third-Order Moments

The third-order moments of velocity fluctuations offer relevant probabilistic information about the flux and the advection of Reynolds normal stresses. In addition, they give an indication of the predominance of turbulent bursting events [26]. The third-order moments, in the generalized form in xz plane, is expressed as $m_{jk} = \overline{\widetilde{u}^j \widetilde{w}^k}$, where $\widetilde{u} = u'/(\overline{u'u'})^{0.5}$, $\widetilde{w} = w'/(\overline{w'w'})^{0.5}$ and $j + k = 3$. Therefore, depending on the values of j and k, the third-order moments are given as, $m_{30} = \overline{u'u'u'}/(\overline{u'u'})^{1.5}$, $m_{03} = \overline{w'w'w'}/(\overline{w'w'})^{1.5}$, $m_{21} = \overline{u'u'w'}/[(\overline{u'u'}) \times (\overline{w'w'})^{0.5}]$ and $m_{12} = \overline{u'w'w'}/[(\overline{u'u'})^{0.5} \times (\overline{w'w'})]$. Here, the m_{30} signifies the skewness of u', indicating the streamwise flux of the streamwise Reynolds normal stress $\overline{u'u'}$. The m_{03} defines the skewness of w', suggesting the vertical flux of the vertical Reynolds normal stress $\overline{w'w'}$. In addition, the m_{21} represents the advection of $\overline{u'u'}$ in the vertical direction, whereas the m_{12} demonstrates the advection of $\overline{w'w'}$ in the streamwise direction.

Figure 7 shows the vertical profiles of m_{30} and m_{03} at the upstream and various downstream relative streamwise distances x/L_d in Runs 1 and 2. Upstream of the dune ($x/L_d = -0.5$), the m_{30} and m_{03}, in the near-bed flow zone, are negative and positive, respectively. Then, they increase with an increase in relative vertical distance z/H_d without changing their signs. Downstream of the dune ($x/L_d = 1$ to 2.1), for a given x/L_d, the m_{30} and m_{03}, in the near-bed flow zone, start with positive and negative values, respectively. Thereafter, they increase slowly with an increase in z/H_d until they attain their respective positive and negative peaks at $z/H_d \approx 0.75$ and 0.5. As the z/H_d increases further, the m_{30} and m_{03} reduce quickly, changing their signs at $z/H_d = 1$, and for $z/H_d > 1$, they become independent of z/H_d. However, these features disappear gradually with an increase in x/L_d. It may be noted that the m_{30} and m_{03} profiles at $x/L_d = 3.3$ remain almost similar to those in the upstream.

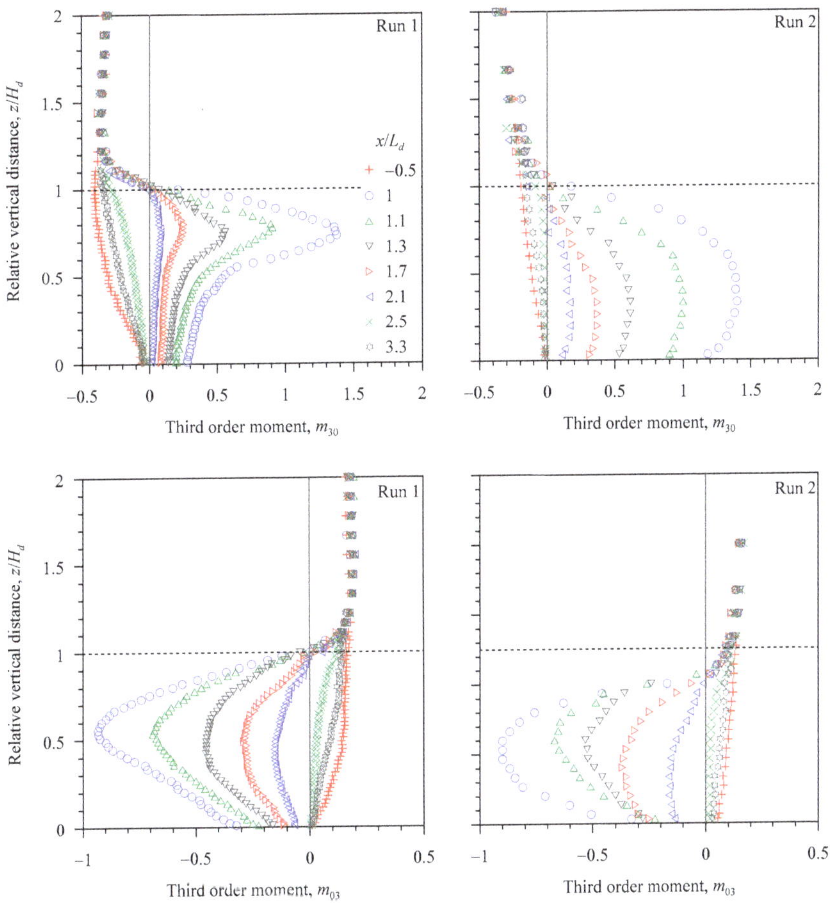

Figure 7. Vertical profiles of third-order moments m_{30} and m_{03} at various relative streamwise distances x/L_d in Runs 1 and 2.

Figure 8 depicts the vertical profiles of m_{21} and m_{12} at the upstream and various downstream relative streamwise distances x/L_d in Runs 1 and 2. It appears that upstream of the dune ($x/L_d = -0.5$), the m_{21} and m_{12}, in the near-bed flow zone, attain positive and negative values, respectively. Then, they increase with an increase in relative vertical distance z/H_d up to a certain height. Subsequently, they reduce with an increase in z/H_d, becoming independent of z/H_d for $z/H_d > 1.1$. Downstream of the dune ($x/L_d = 1$ to 2.1), for a given x/L_d, the m_{21} and m_{12}, in the near-bed flow zone, are negative and positive, respectively. Then, they increase with an increase in z/H_d attaining their respective peaks. Afterward, they reduce quickly, changing their signs at the dune crest ($z/H_d = 1$). Thereafter, the m_{21} and m_{12} profiles recover their upstream profiles. Downstream of the dune, an advection of $\overline{u'u'}$ in the upward direction and that of $\overline{w'w'}$ in the upstream direction prevail below the crest. In fact, below the crest, there appears a streamwise acceleration, which is linked with the downward flux causing sweeps with an advection of $\overline{u'u'}$ in the downward direction. By contrast, above the crest, the streamwise deceleration is associated with an upward flux producing ejections with an advection of $\overline{u'u'}$ in the upward direction.

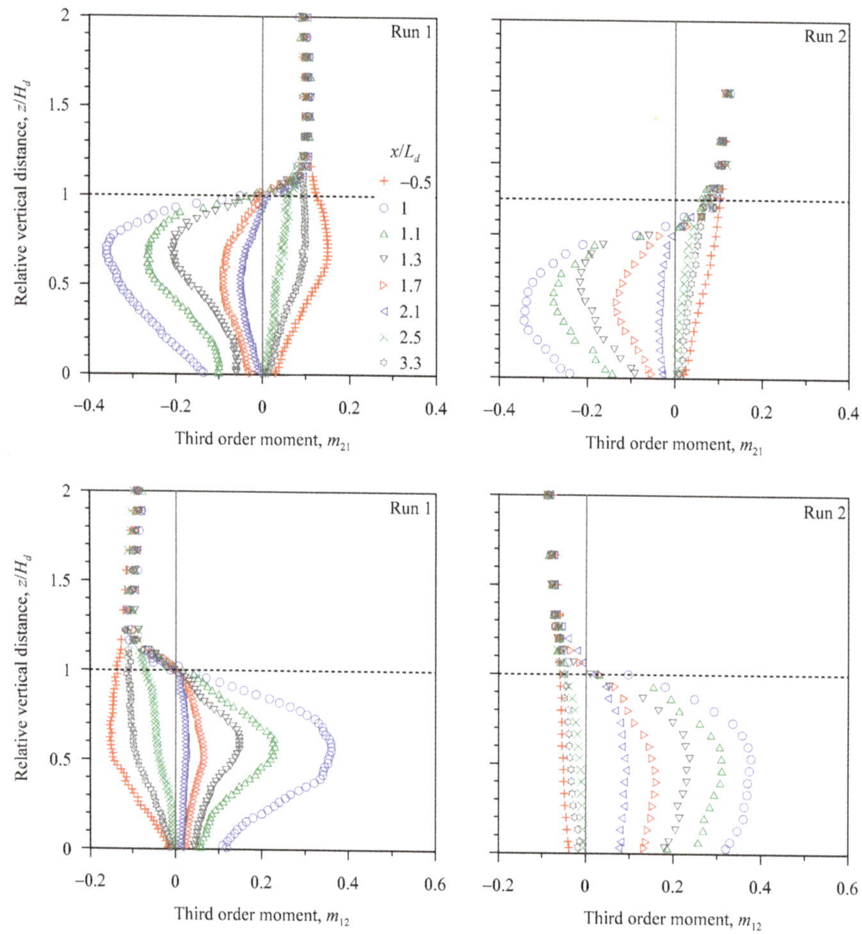

Figure 8. Vertical profiles of third-order moments m_{21} and m_{12} at various relative streamwise distances x/L_d in Runs 1 and 2.

5. Quadrant Analysis

Lu and Willmarth [27] suggested that the bursting events can be quantified by performing the quadrant analysis of velocity fluctuations u' and w' on a $u'w'$ plane. The turbulent bursting includes four events in four distinct quadrants $i = 1$ to 4, such as (i) Q1 events or *outward interactions* ($i = 1$ and u', $w' > 0$), (ii) Q2 events or *ejections* ($i = 2$ and $u' < 0$, $w' > 0$), (iii) Q3 events or *inward interactions* ($i = 3$ and u', $w' < 0$) and (iv) Q4 events or *sweeps* ($i = 4$ and $u' > 0$, $w' < 0$). Outside the *hole size* H, the contribution of $\overline{u'w'}\big|_{i,H}$ from the quadrant i to $\overline{u'w'}$ is ascertained by averaging the quantity $u'(t)w'(t)F_{i,H}$ over the sampling duration. Here, $F_{i,H}$ is the *detection function*, defined as $F_{i,H} = 1$ if the pair (u', w') in the quadrant i satisfies the condition $|u'w'| \geq H(\overline{u'u'})^{0.5}(\overline{w'w'})^{0.5}$ and $F_{i,H} = 0$ otherwise. The *relative fractional contributions* $S_{i,H}$ toward the Reynolds shear stress production is expressed as $S_{i,H} = \overline{u'w'}\big|_{i,H}/\overline{u'w'}$. It turns out that for $H = 0$, the sum of $S_{1,0}$, $S_{2,0}$, $S_{3,0}$ and $S_{4,0}$ becomes unity.

Figures 9 and 10 show the vertical profiles of $S_{i,0}$ at the upstream and various downstream relative streamwise distances x/L_d in Runs 1 and 2, respectively. Upstream of the dune ($x/L_d = -0.5$), the $Q2$ and $Q4$ events remain the most and the second-most contributing events, respectively, to the production of Reynolds shear stress. However, the $Q1$ and $Q3$ events are trivial across the flow depth. Downstream of the dune ($x/L_d = 1$ to 2.1), all the four events contribute largely below the dune crest with prevailing $Q4$ events in the form of arrival of high-speed fluid streaks. At $x/L_d = 2.5$, contributions from the $Q2$ and $Q4$ events appear to be nearly equal below the crest. Further downstream ($x/L_d = 3.3$), the $Q2$ events dominate over $Q4$ events in the form of arrival of low-speed fluid streaks. It may be noted that above the crest ($z/H_d > 1$), the $Q2$ events are the most contributing events regardless of x/L_d.

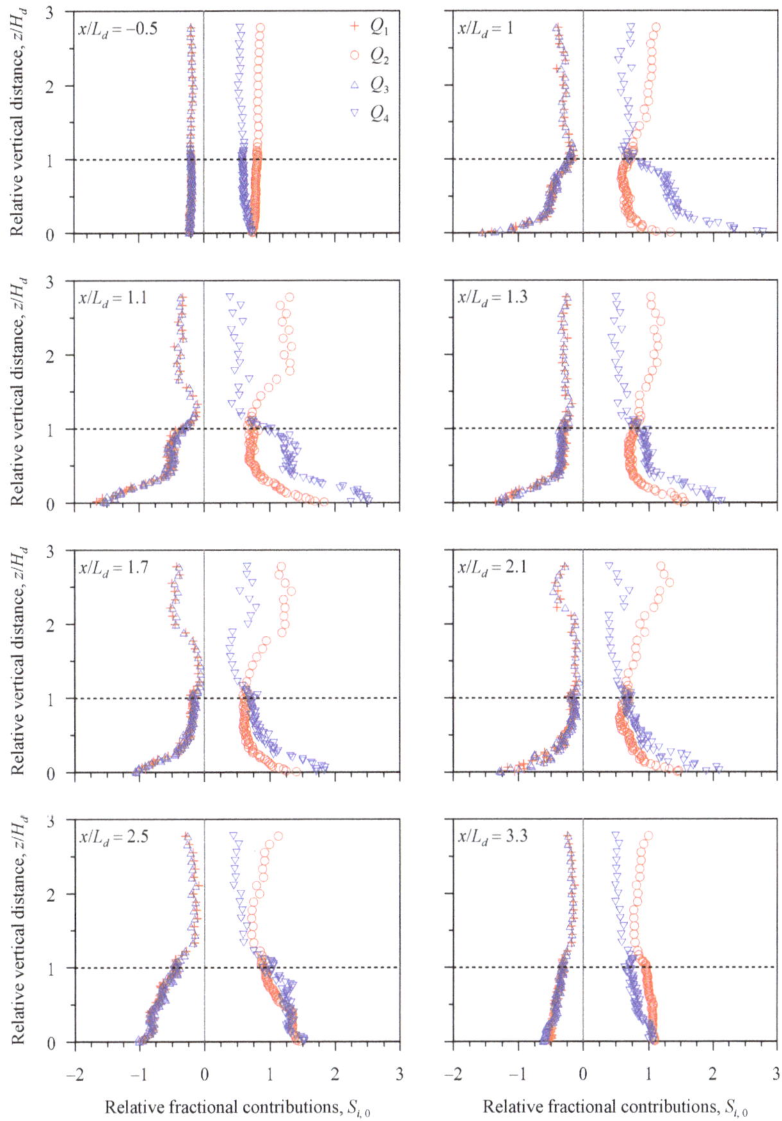

Figure 9. Vertical profiles of relative fractional contributions $S_{i,0}$ at various relative streamwise distances x/L_d in Run 1.

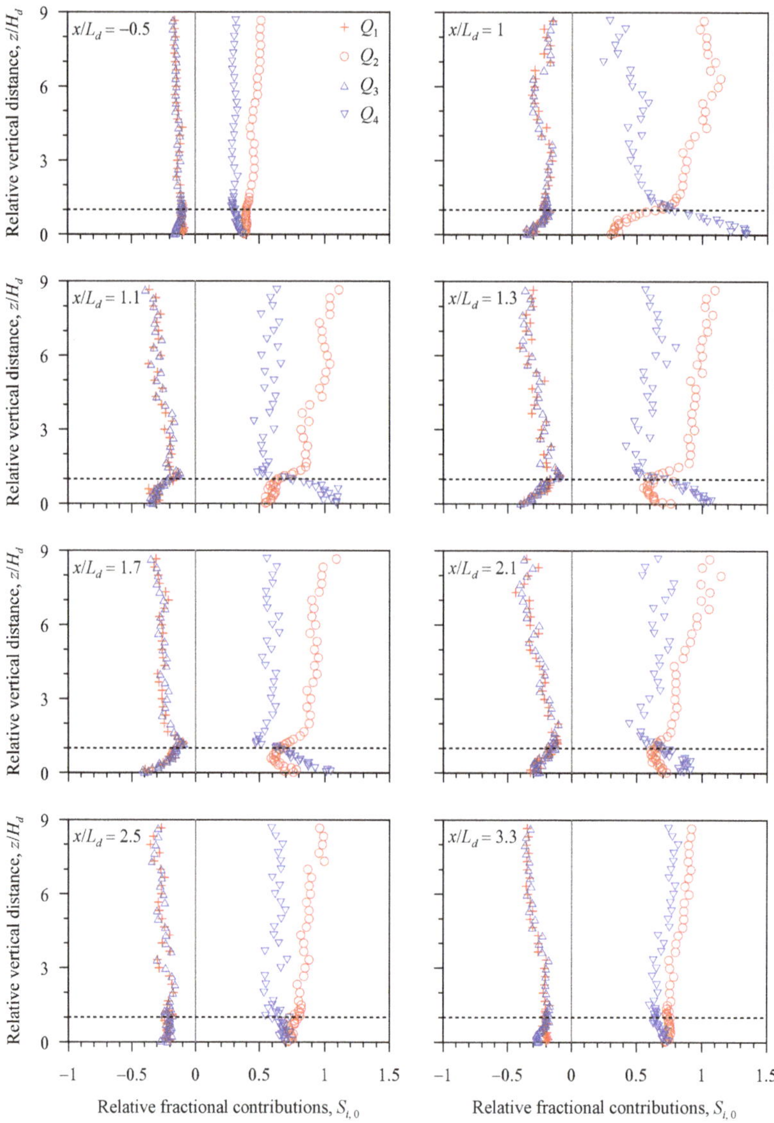

Figure 10. Vertical profiles of relative fractional contributions $S_{i,0}$ at various relative streamwise distances x/L_d in Run 2.

Figure 11a,b shows the variations of relative fractional contributions $|S_{i,H}|$ with hole size H in Run 1 for different relative vertical distances z/H_d (=0.05, 0.25 and 0.5) at relative streamwise distances $x/L_d = -0.5$ (uninterrupted upstream flow) and 1 (near-wake flow), whereas Figure 12a,b shows those at $x/L_d = 1.7$ (far-wake flow) and 3.3 (near to fully recovered flow). It appears that upstream of the dune ($x/L_d = -0.5$), the Q1 and Q3 events for $z/H_d = 0.05$ contribute minimally to the Reynolds shear stress production as compared to the Q2 and Q4 events. However, for $z/H_d = 0.05$, the pairs (Q1, Q3) and (Q2, Q4) are equal, indicating that they mutually cancel the dominance of each other. At $x/L_d = -0.5$, the Q2 events remain dominant for $z/D = 0.25$ and 0.5. Immediate downstream of the dune ($x/L_d = 1$), the Q1 and Q3 events, for a given z/H_d, are smaller than Q2 and Q4 events. However, at the downstream, the $Q4$ remain the most dominant events for $z/H_d = 0.05$, 0.25 and 0.5. At $x/L_d = 1.7$, these features remain similar to those at $x/L_d = 1$, but with relatively smaller $Q4$ events. Far downstream of the dune ($x/L_d = 3.3$), the events, for a given z/H_d, follow the upstream trend. The contributions from the events are considerable for lower values of H. In essence, for $H \geq 12$, all the events are trivial at different streamwise and vertical distances.

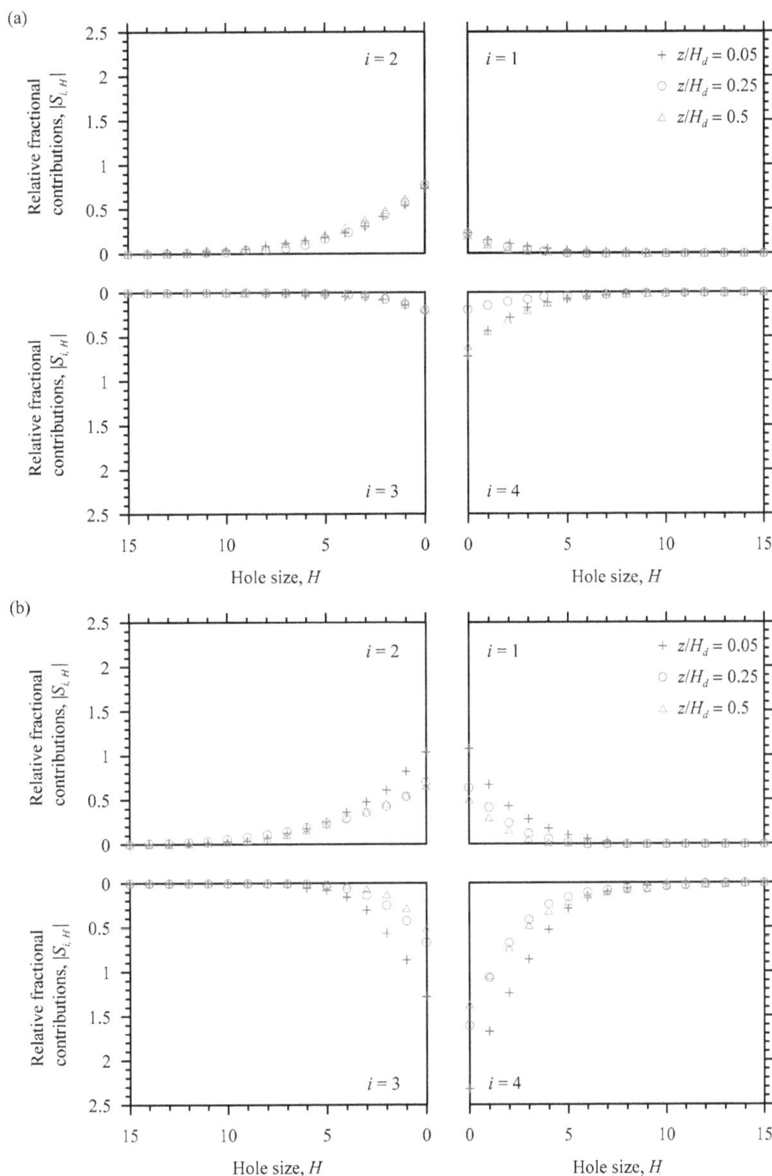

Figure 11. Relative fractional contributions $|S_{i,H}|$ as a function of hole size H in Run 1 at relative streamwise distances (a) $x/L_d = -0.5$ and (b) $x/L_d = 1$ for relative vertical distances $z/H_d = 0.05$, 0.25 and 0.5.

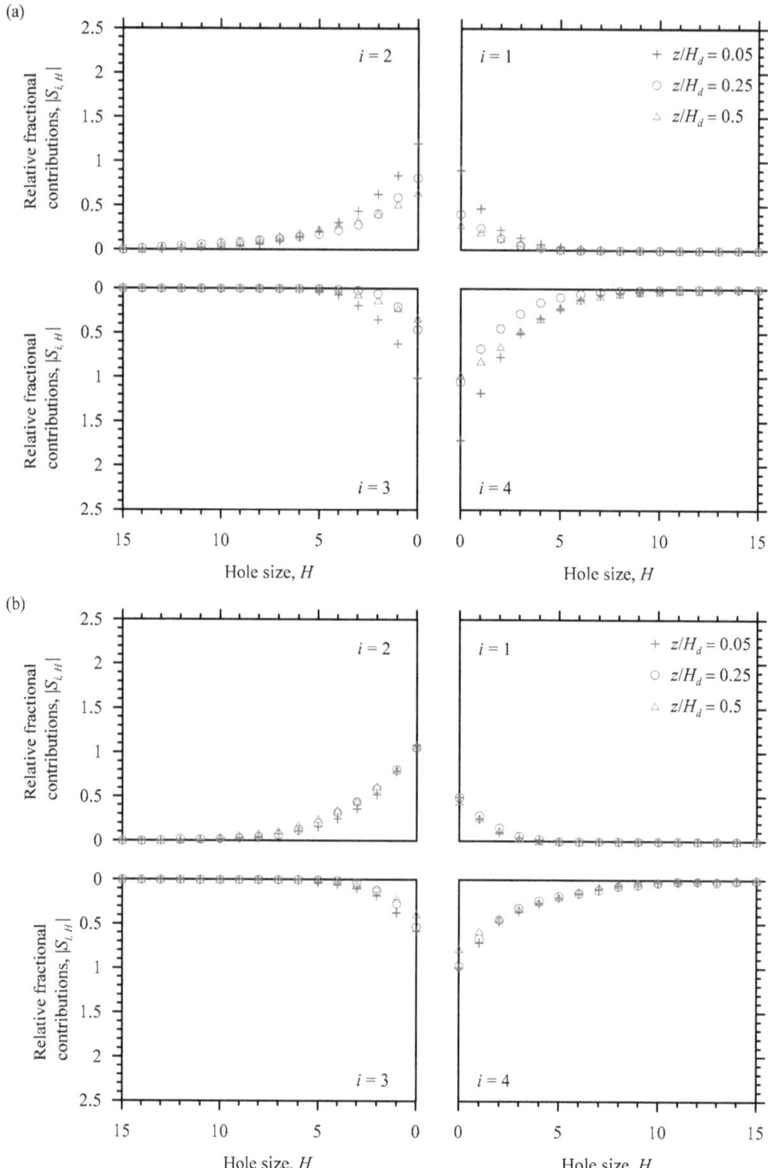

Figure 12. Relative fractional contributions $|S_{i,H}|$ as a function of hole size H in Run 1 at relative streamwise distances (a) $x/L_d = 1.7$ and (b) $x/L_d = 3.3$ for relative vertical distances $z/H_d = 0.05$, 0.25 and 0.5.

6. Turbulent Kinetic Energy Budget

The turbulent kinetic energy budget reads $t_P = \varepsilon + t_D + p_D - v_D$, where t_P is the turbulent kinetic energy production rate ($= -\overline{u'w'}\partial\bar{u}/\partial z$), ε is the turbulent kinetic energy dissipation rate, t_D is the turbulent kinetic energy diffusion rate ($= \partial f_{kw}/\partial z$), f_{kw} is the vertical flux of turbulent kinetic energy, p_D is the pressure energy diffusion rate [$= \rho^{-1}\partial(\overline{p'w'})/\partial z$], p' is the pressure fluctuations, v_D is the viscous diffusion rate ($= v\partial^2 k/\partial z^2$) and k is the turbulent kinetic energy. In an open channel flow, the v_D

is insignificant compared to other components of the turbulent kinetic energy budget. In this study, Kolmogorov second hypothesis was applied to determine the ε from the velocity power spectra [28]. The t_P and t_D were determined from the experimental data, whereas the p_D was obtained from the relationship $p_D = t_P - \varepsilon - t_D$. In nondimensional form, the set of variables $(t_P, \varepsilon, t_D, p_D)$ is expressed as $(T_P, E_D, T_D, P_D) = (t_P, \varepsilon, t_D, p_D) \times (H_d/u_*^3)$.

Figure 13 illustrates the vertical profiles of nondimensional components of the turbulent kinetic energy budget at various relative streamwise distances x/L_d in Run 1. Upstream of the dune ($x/L_d = -0.5$), all the components of the turbulent kinetic energy budget, in the near-bed flow zone, are positive with a sequence of magnitude $T_P > E_D > P_D > T_D$ and then, they reduce with an increase in relative vertical distance z/H_d. Above the dune crest ($z/H_d > 1$), they are quite small. Downstream of the dune ($x/L_d = 1$ to 2.1), the peaks of T_P, E_D, P_D and T_D are found to appear at the crest. In the near-bed flow zone, the T_P and E_D are positive, whereas the P_D and T_D are negative for $x/L_d = 1$ to 2.1. Downstream of the dune, the absolute values of T_P, E_D, P_D and T_D decrease with an increase in x/L_d. In particular, at $x/L_d = 3.3$, the T_P, E_D, P_D and T_D profiles are almost similar to those of the undisturbed upstream flow at $x/L_d = -0.5$.

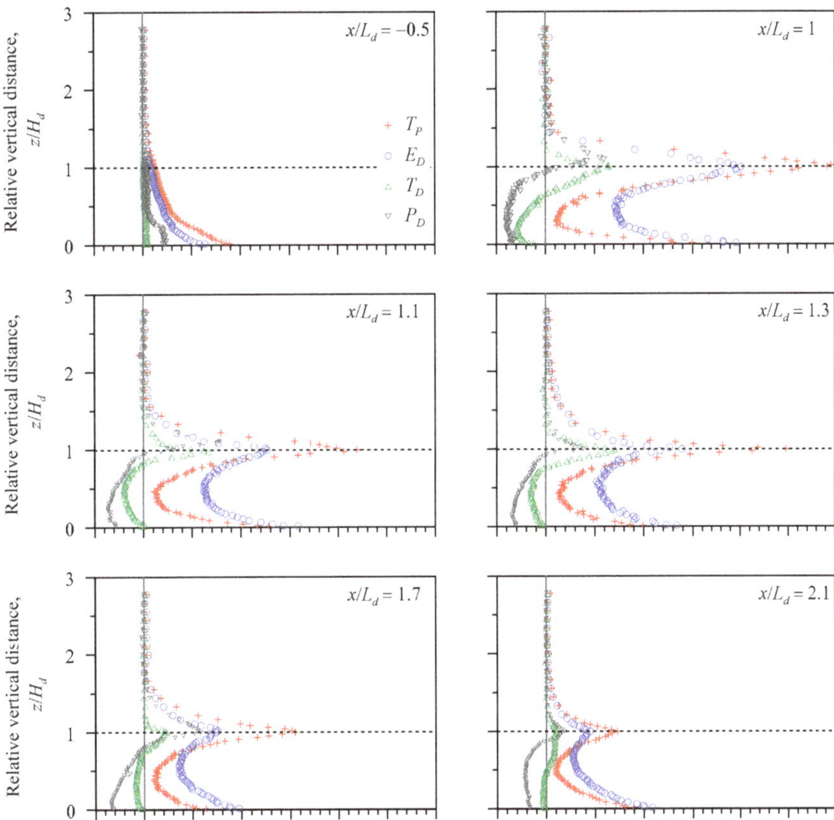

Figure 13. *Cont.*

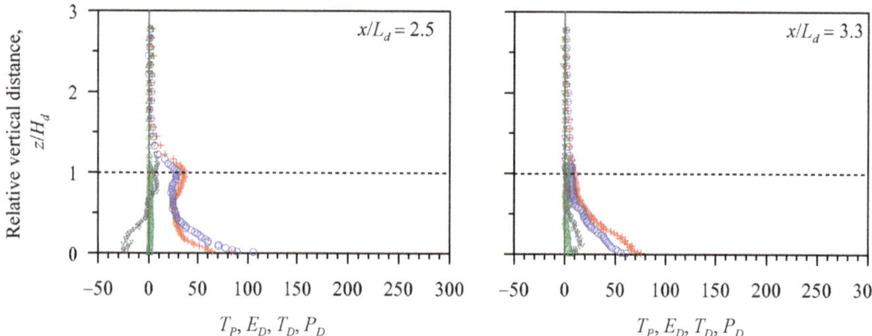

Figure 13. Vertical profiles of the nondimensional components of turbulent kinetic energy budget T_P, E_D, T_D and P_D at various relative streamwise distances x/L_d in Run 1.

7. Reynolds Stress Anisotropy

An *isotropic turbulence* refers to an idealized condition, where the velocity fluctuations at a specific point remain invariant to the rotation of axes. In a lucid way, this condition indicates that the Reynolds normal stresses are identical ($\sigma_x = \sigma_y = \sigma_z$), where ($\sigma_x$, σ_y, σ_z) = ($\overline{u'u'}$, $\overline{v'v'}$, $\overline{w'w'}$). By contrast, in an *anisotropic turbulence*, the Reynolds normal stresses are dissimilar, because the velocity fluctuations u'_i [= (u', v', w') for i = (1, 2, 3)] are directionally preferred.

The *Reynolds stress anisotropy tensor* b_{ij} is expressed as $b_{ij} = \overline{u'_i u'_j}/(2k) - \delta_{ij}/3$, where δ_{ij} is the Kronecker delta function [$\delta_{ij}(i = j)$ = 1 and $\delta_{ij}(i \neq j)$ = 0]. To ascertain the degree and the nature of anisotropy, the second and third principal invariants, I_2 (= $-b_{ij}b_{ij}/2$) and I_3 (= $b_{ij}b_{jk}b_{ki}/3$), respectively, are introduced. The Reynolds stress anisotropy is determined by plotting $-I_2$ as a function of I_3, called the *anisotropy invariant map* (AIM). In an AIM, the possible turbulence states are confined to a triangle, called the *Lumley triangle* (Figure 14). The left-curved and the right-curved boundaries of the Lumley triangle, given by $I_3 = \pm 2(-I_2/3)^{3/2}$, are symmetric about the *plane-strain limit* (I_3 = 0). In addition, the top-linear boundary of the Lumley triangle obeys $I_3 = -(9I_2 + 1)/27$. Dey et al. [29] envisioned the Reynolds stress anisotropy from the perspective of the shape of ellipsoid formed by the Reynolds normal stresses (σ_x, σ_y, σ_z) in (x, y, z). In an isotropic turbulence ($\sigma_x = \sigma_y = \sigma_z$), the stress ellipsoid becomes a *sphere* (Figure 14). On the left-curved boundary, called the *axisymmetric contraction limit*, one component of Reynolds normal stress is smaller than the other two equal components ($\sigma_x = \sigma_y > \sigma_z$), forming the stress ellipsoid an *oblate spheroid*. On the left vertex, called the *two-component axisymmetric limit*, one component of Reynolds normal stress disappears ($\sigma_x = \sigma_y$ and σ_z = 0) to make the stress ellipsoid a *circular disc* (Figure 14). On the right-curved boundary, called the *axisymmetric expansion limit*, one component of Reynolds normal stress is larger than the other two equal components ($\sigma_x = \sigma_y < \sigma_z$), making the stress ellipsoid a *prolate spheroid* (Figure 14). Further, on the top-linear boundary, called the *two-component limit*, one component of Reynolds normal stress is larger than the other component together with a third vanishing component ($\sigma_x > \sigma_y$ and σ_z = 0), producing the stress ellipsoid an *elliptical disk*. The point of intersecting of the plain-strain limit and the two-component limit is called the *two-component plain-strain limit*. Moreover, on the right vertex of the Lumley triangle, called the *one-component limit* [($\sigma_x > 0$, $\sigma_y = \sigma_z = 0$) or ($\sigma_x = \sigma_y = 0$, $\sigma_z > 0$)], only one component of Reynolds normal stress sustains to make the stress ellipsoid a *straight line* (Figure 14).

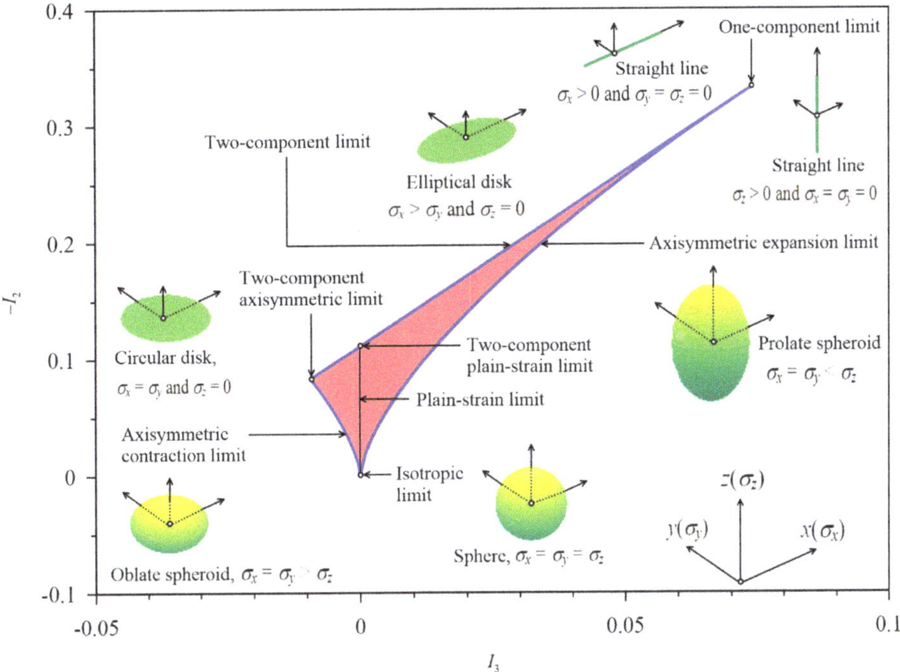

Figure 14. Conceptual representation of Reynolds stress anisotropy.

Figure 15 shows the data plots of $-I_2$ versus I_3, confined to the AIM boundaries, at various relative streamwise distances x/L_d in Runs 1 and 2. Upstream of the dune ($x/L_d = -0.5$), the data plots initiate from the near left vertex, moving toward the bottom cusp, and then, with an increase in vertical distance, they cross the plain-strain limit to shift toward the right-curved boundary. The trends of the data plots for both Runs 1 and 2 are almost monotonic. The AIM of the upstream indicates that as the vertical distance increases, the turbulence anisotropy tends to reduce to a quasi-three-dimensional isotropy. Immediate downstream of the dune, the data plots tend to create a stretched loop inclined to the left-curved boundary. However, below the dune crest ($z/H_d < 1$), the data plots in the near-bed flow zone initiate from the plain-strain limit and with an increase in vertical distance up to the crest, they shift toward the left vertex following the left-curved boundary. This suggests that the turbulence anisotropy has an affinity to a two-dimensional isotropy. Above the crest, the data plots turn toward the right and as the vertical distance increases further, they move toward the bottom cusp following the left-curved boundary. This demonstrates that the turbulence anisotropy tends to reduce to a quasi-three-dimensional isotropy. Further downstream ($x/L_d = 1.7$), the size of the loop created by the data plots reduces forming a tail, and the loop disappears at $x/L_d = 3.3$, signifying a recovery of the undisturbed upstream trend. It therefore appears that that below the crest, the turbulence has an affinity to a two-dimensional isotropy, whereas above the crest, a quasi-three-dimensional isotropy prevails.

From the perspective of the shape of stress ellipsoid, Figure 15 shows that below the dune crest, an oblate spheroid axisymmetric turbulence is predominant in the wall-wake flow. The line of plain-strain limit ($I_3 = 0$) is touched by the curve through the data plots in the near-bed flow zone. This reveals that the axisymmetric contraction to the oblate spheroid enhances as the vertical distance increases up to the crest. However, the axisymmetric contraction to oblate spheroid lessens with a further increase in vertical distance above the crest.

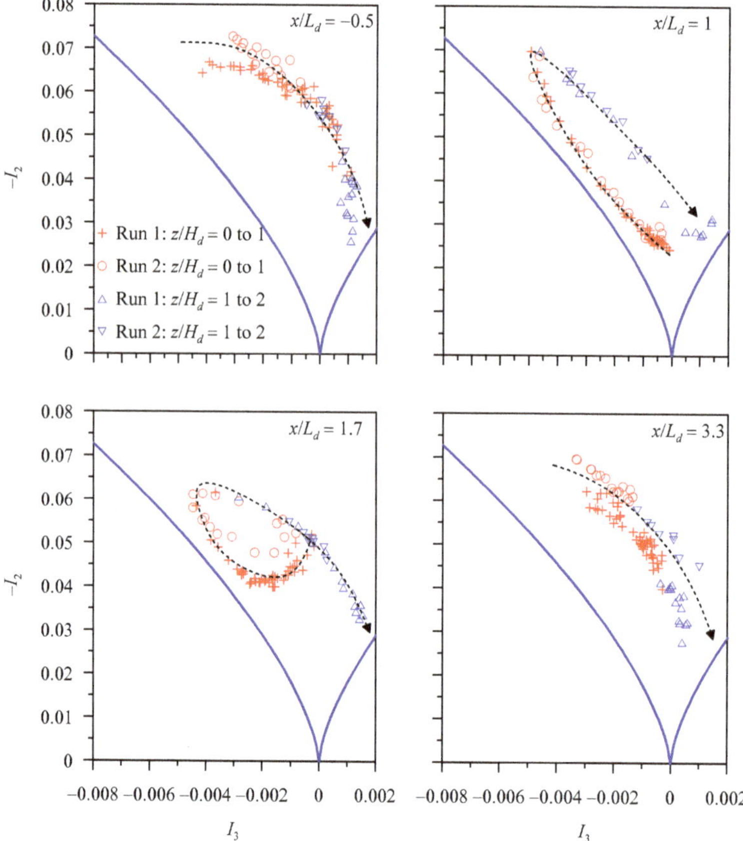

Figure 15. AIMs at various relative streamwise distances x/L_d in Runs 1 and 2.

8. Conclusions

This study puts into focus the turbulence in wall-wake flow downstream of an isolated dunal bedform. The vertical profiles of streamwise flow velocity reveal that the near-wake flow zone extends up to 1.7 times the dune length, whereas the streamwise flow velocity profile follows the undisturbed upstream velocity profile beyond 3.3 times the dune length. The Reynolds shear stress in the wall-wake flow is affected by the dune up to a vertical distance of 1.75 times the dune height and a streamwise distance of 2.5 times the dune length. The third-order moment of velocity fluctuations reveal that downstream of the dune, a streamwise acceleration having a downward flux prevails below the dune crest, whereas a streamwise deceleration having an upward flux persists above the crest. Below the crest, the sweeps are found to be the predominant events, whereas above the crest, the ejections are the major events. The components of the turbulent kinetic energy budget reveal an amplification of the magnitudes of the turbulent parameters, which attain their maximum peaks at the crest. The anisotropy invariant maps show that the data plots in the wall-wake flow start from the plain-strain limit in the near-bed flow zone, shifting toward the left vertex of the Lumley triangle up to the crest to show an affinity to a two-dimensional isotropy. Above the crest, the data plots show an affinity to a quasi-three-dimensional isotropy.

In essence, this study advances the current understanding of flow and turbulence characteristics in wall-wake flow downstream of an isolated dunal bedform. The experimental results provide guidance to numerical simulations of wall-wake flow. In addition, this study may be helpful, at least qualitatively, to simulate the mobile-bed flow downstream of a dunal bedform.

Author Contributions: Conceptualization, S.S. and S.D; data curation, S.S.; formal analysis, S.S., S.Z.A. and S.D; funding acquisition, S.S.; investigation, S.S., S.Z.A. and S.D; methodology, S.S. and S.D; resources, S.S., S.Z.A. and S.D; writing—original draft, S.S., S.Z.A. and S.D; writing—review and editing, S.S., S.Z.A. and S.D; supervision, S.D.

Funding: This research was funded by Indian Statistical Institute, Kolkata.

Acknowledgments: The third author (S.D) acknowledges the JC Bose fellowship project (project code: JBD) to coordinate this research program.

Conflicts of Interest: The authors declare no conflict of interest.

References

1. ASCE Task Committee. Flow and transport over dunes. *J. Hydraul. Eng.* **2002**, *128*, 726–728. [CrossRef]
2. McLean, S.R.; Smith, J.D. A model for flow over two-dimensional bed forms. *J. Hydraul. Eng.* **1986**, *112*, 300–317. [CrossRef]
3. Nelson, J.M.; Smith, J.D. Mechanics of flow over ripples and dunes. *J. Geophys. Res. Oceans* **1989**, *94*, 8146–8162. [CrossRef]
4. Maddux, T.B.; Nelson, J.M.; McLean, S.R. Turbulent flow over three-dimensional dunes: 1. Free surface and flow response. *J. Geophys. Res. Earth Surf.* **2003**, *108*, 6009. [CrossRef]
5. Maddux, T.B.; McLean, S.R.; Nelson, J.M. Turbulent flow over three-dimensional dunes: 2. Fluid and bed stresses. *J. Geophys. Res. Earth Surf.* **2003**, *108*, 6010. [CrossRef]
6. Sukhodolov, A.N.; Fedele, J.J.; Rhoads, B.L. Structure of flow over alluvial bedforms: An experiment on linking field and laboratory methods. *Earth Surf. Proc. Landf.* **2006**, *31*, 1292–1310. [CrossRef]
7. Best, J.L. Kinematics, topology and significance of dune-related macroturbulence: Some observations from the laboratory and field. In *Fluvial Sedimentology VII*; Blum, M.D., Marriott, S.B., Leclair, S.F., Eds.; Special Publication of International Association of Sedimentologists, Wiley-Blackwell (an imprint of John Wiley & Sons Ltd): Hoboken, NJ, USA, 2005; pp. 41–60.
8. Schlichting, H. *Boundary Layer Theory*; McGraw-Hill: New York, NY, USA, 1979.
9. Balachandar, R.; Ramachandran, S.; Tachie, M.F. Characteristics of shallow turbulent near wakes at low Reynolds numbers. *J. Fluids Eng.* **2000**, *122*, 302–308. [CrossRef]
10. Tachie, M.F.; Balachandar, R. Shallow wakes generated on smooth and rough surfaces. *Exp. Fluids* **2001**, *30*, 467–474. [CrossRef]
11. Shamloo, H.; Rajaratnam, N.; Katopodis, C. Hydraulics of simple habitat structures. *J. Hydraul. Res.* **2001**, *39*, 351–366. [CrossRef]
12. Dey, S.; Sarkar, S.; Bose, S.K.; Tait, S.; Castro-Orgaz, O. Wall-wake flows downstream of a sphere placed on a plane rough wall. *J. Hydraul. Eng.* **2011**, *137*, 1173–1189. [CrossRef]
13. Sarkar, S.; Dey, S. Turbulent length scales and anisotropy downstream of a wall mounted sphere. *J. Hydraul. Res.* **2015**, *53*, 649–658. [CrossRef]
14. Kahraman, A.; Sahin, B.; Rockwell, D. Control of vortex formation from a vertical cylinder in shallow water: Effect of localized roughness elements. *Exp. Fluids* **2002**, *33*, 54–65. [CrossRef]
15. Akilli, H.; Rockwell, D. Vortex formation from a cylinder in shallow water. *Phys. Fluids* **2002**, *14*, 2957–2967. [CrossRef]
16. Ozgoren, M. Flow structure in the downstream of square and circular cylinders. *Flow Meas. Instrum.* **2006**, *17*, 225–235. [CrossRef]
17. Ozturk, N.A.; Akkoca, A.; Sahin, B. Flow details of a circular cylinder mounted on a flat plate. *J. Hydraul. Res.* **2008**, *46*, 344–355. [CrossRef]
18. Sadeque, M.A.F.; Rajaratnam, N.; Loewen, M.R. Shallow turbulent wakes behind bed-mounted cylinders in open channels. *J. Hydraul. Res.* **2009**, *47*, 727–743. [CrossRef]
19. Dey, S.; Swargiary, D.; Sarkar, S.; Fang, H.; Gaudio, R. Self-similarity in turbulent wall-wake flow downstream of a wall-mounted vertical cylinder. *J. Hydraul. Eng.* **2018**, *144*, 04018023. [CrossRef]

20. Dey, S.; Swargiary, D.; Sarkar, S.; Fang, H.; Gaudio, R. Turbulence features in a wall-wake flow downstream of a wall-mounted vertical cylinder. *Eur. J. Mech. B Fluids* **2018**, *69*, 46–61. [CrossRef]
21. Dey, S.; Lodh, R.; Sarkar, S. Turbulence characteristics in wall-wake flows downstream of wall-mounted and near-wall horizontal cylinders. *Environ. Fluid Mech.* **2018**, *18*, 891–921. [CrossRef]
22. Lacey, R.W.J.; Roy, A.G. Fine-scale characterization of the turbulent shear layer of an instream pebble cluster. *J. Hydraul. Eng.* **2008**, *134*, 925–936. [CrossRef]
23. Sarkar, S.; Dey, S. Turbulence in wall-wake flow downstream of an isolated dune. In Proceedings of the 38th International School of Hydraulics, Łąck, Poland, 21–24 May 2019.
24. Dey, S.; Sarkar, S. Turbulent length scales and Reynolds stress anisotropy in wall-wake flow downstream of an isolated dunal bedform. In Proceedings of the 38th International School of Hydraulics, Łąck, Poland, 21–24 May 2019.
25. Goring, D.G.; Nikora, V.I. Despiking acoustic Doppler velocimeter data. *J. Hydraul. Eng.* **2002**, *128*, 117–126. [CrossRef]
26. Gad-El-Hak, M.; Bandyopadhyay, P.R. Reynolds number effects in wall-bounded turbulent flows. *Appl. Mech. Rev.* **1994**, *47*, 307–365. [CrossRef]
27. Lu, S.S.; Willmarth, W.W. Measurements of the structure of the Reynolds stress in a turbulent boundary layer. *J. Fluid Mech.* **1973**, *60*, 481–511. [CrossRef]
28. Dey, S. *Fluvial Hydrodynamics: Hydrodynamic and Sediment. Transport. Phenomena*; Springer: Berlin, Germany, 2014.
29. Dey, S.; Ravi Kishore, G.; Castro-Orgaz, O.; Ali, S.Z. Turbulent length scales and anisotropy in submerged turbulent plane offset jets. *J. Hydraul. Eng.* **2019**, *145*, 04018085. [CrossRef]

MDPI
St. Alban-Anlage 66
4052 Basel
Switzerland
Tel. +41 61 683 77 34
Fax +41 61 302 89 18
www.mdpi.com

Water Editorial Office
E-mail: water@mdpi.com
www.mdpi.com/journal/water